Manufacturing Technology-I

Dr.R. Sivasubramanian, B.E., M.E., Ph.D.,

Principal,

Latha Mathavan Engineering College,

Kidaripatti, Madurai.

Published by

Second Edition: March 2019

ISBN 978-93-86638-24-3

Author

Dr.R. Sivasubramanian

Bonfring

309, 2nd Floor, 5th Street Extension, Gandhipuram,

Coimbatore-641 012.

Tamilnadu, India.

E-mail: info@bonfring.org

Website: www.bonfring.org

Phone: 0422 4213231

Dedicated

To the loving

Parents, my family and my colleagues and friends

<table>
<tr><th>Unit</th><th>Contents</th><th>Page No</th></tr>
</table>

Unit 1

Casting

1.1. Sand Casting

- It is the one of the process used to making components of complicated shapes.
- It producing metal parts by pouring molten metal into mould cavity of shape.
- The allowing metal solidifies and the metal piece is called after solidification is called casting.

Sand Mould

- It is the cavity of the required shape in moulding sand.

Foundry

- Pattern making
- Casting
- Mould making

1.2. Pattern Making

- Pattern is one of the important tools
- It's making cavity into the mould to produce a casting

1.3. Types of Pattern

- Solid Pattern
- Split Pattern
- Match Plate Pattern
- Loose Piece Pattern
- Sweep Pattern
- Skeleton Pattern
- Shall Pattern
- Segmental Pattern

Tabulation 1.1

Solid pattern	It's used for making few large size casting	Shapes made easily
Split pattern	Symmetrical shaped costing such as Sphere	it's used symmetrical shape
Loose piece pattern	If a pattern is made from a single piece having projection. Its build up into patterns	It's used for large size simple casting
Match plate	Two halves mounted on both sides	It's used machine moulding
Skelton pattern	It's made of wood. Board is remove the excess sand	It's used in pipes, turbine casting
Part pattern	The bottom of the mould is rammed	It's used for gear blanks
Shell pattern	It's an hollow pattern	It's mainly used for drain age

1.4. Pattern Material

- Wood
- Metal
- Wax
- Plastic
- Plaster

Tabulation 1.2

WOOD • Teak wood • Mahogany • White pine	Its commonly used for pattern making. It contain more than 10% moisture avoid wrapping	Accuracy is high. Light in weight. Easily repaired.	Non uniform structure High tear by sand Cannot be used in machine moulding
Metal • Cast iron • Brass • Aluminium	It's used in machine moulding It's used in soldering	Long life Smooth surface Mass production	Cost is high Cannot easily repaired
Plaster	Gypsum cement is used for preparing hollow boards	Easily worked It's used only small patterns	Affected by moisture
Plastics	It does not sink more epical	High dimension accuracy	Light in weight

Pattern Allowances

- It not made into the exact size
- It made larger than the required casting

Types

- Shrinkage allowances
- Finish allowances
- Taber allowances
- Camber allowances
- Rapping allowances

Shrinkage Allowances

Definition

- The metal shrinks on solidification and contracts further on cooling at room temperature.

Materials

- CI -10.4 mm
- AL-17 mm
- Brass-15.3 mm

Finishing Allowance

- All the casting is to be machined to get required on the shape
- The extra size is given to the pattern

Materials

- CI-2.5 mm
- AL-1.6 mm
- BR-1.6 mm

Taper Allowance

- If the vertical faces of pattern are perpendicular to parting line.
- The edges mould may be damaged
- When the pattern is removed from the sand
- Tapper is easy removal of pattern
- It's called taper allowance.

Diagram

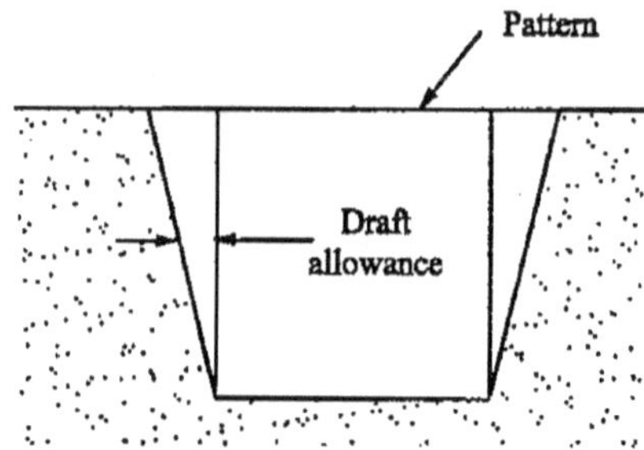

Figure 1.1: Draft Allowances

Camber Allowance

- The casting may be distort
- It is an irregular shape
- To avoid the shape of the pattern is opposite direction

Shake Allowances

- To remove the pattern mould cavity of slightly rapped
- It's also called negative allowances.
- It may have several projections
- It also has extra projection to produce the runners

Properties of Moulding Sand

- Porosity
- plasticity
- Adhesiveness
- Strength
- Refractoriness
- collapse/ability

Porosity

- It's a measure of moulding sand by which the sand allows the steam to passes through it.
- When molten metal is poured into the mould steam are formed due to moisture
- If the steams are not removed, blow holes will not occur.
- Parameters affect the porosity

- Day content is less-porosity more
- Grain size larger –permeability more

Plasticity

- It is the property of moulding sand by which the moulding around and over the pattern fills in flasks.
- The property may be improved by adding water.

Adhesiveness

- Which it sticks to another body
- It depends on the type and amount of builder used in sand mix.

Strength

- Its property of moulding sand by which sticks together.
- Moulding sand should have sufficient strength so mould does not collapse.
- The sand depends upon grain size.

Refractoriness

- It resists high temperature of molten metal.
- The property depends upon purity of sand.

Collapsibility

- The property of moulding sand to decrease in some forces enveloped by shrinkage of metal.
- If the mould does not collapse.
- It depends upon amount of quartz.

Testing of Moulding Sand

- The moulding sands have to be correct in size and good surface.
- It contains the silica sand, clay moisture content.
- Finally the sound castings are produced

1.5. Types of Moulding Sand Machine Testing

- Moisture content test
- Clay content test
- Strength test
- Grain fitness test

- Porosity test
- Tough test
- Hot strength test
- Mould hardness test

1.5.1. *Moisture Content Test*

- Introduction
- Types
- Diagram
- Construction
- Working
- Merits
- Demerits

Introduction

- It plays a very important role in moulding sand.

Types

- Loss of weight after evaporation
- Moisture teller method
- Moisture based on chemical reaction
- Loss of weight after evaporation

Diagram

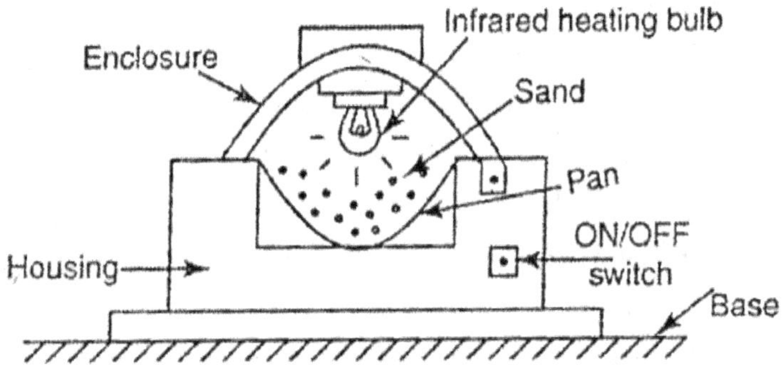

Figure 1.2: Moisture Content Method

Construction

The heating bulb is connected

- The enclosure is hinged at one end but it's freely supported at other end.
- The heating bulb is off by operating two way switches.
- Entire housing is mounted on the base.

Working

- Measured quantity of sample is taken
- It is powered into the pan provided on the apparatus.
- Pan is closed by enclosure.
- Heating bulb is switched on when emits heating ray towards pan.
- Heating is done for 3 mins
- When moulding is evaporated
- Supply the bulb is switch off
- Then it's reweighted subtracting the final weight from original weight.

Merits

- Simple process
- Cost is low

Demerits

- Time consumption is more

1.5.2. Moisture Teller Method

- Diagram
- Construction and Working
- Merits
- Demerits

Diagram

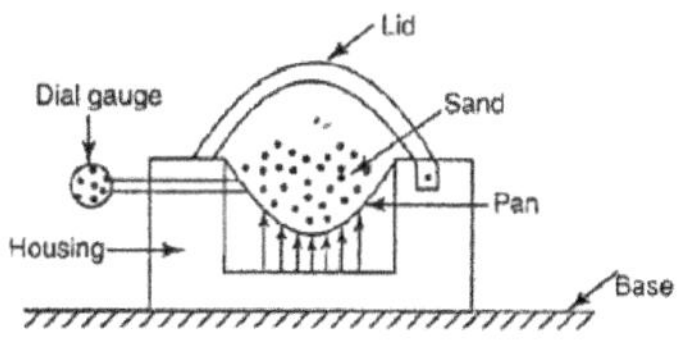

Figure 1.3: Moisture Teller Method

Construction and Working

- On the top of the housing an enclosure the heating bulb is connected
- Housing is mounted on the base
- The weighted quantity of sand sample is taken
- The pan has about 500 mesh screen in the bottom
- The pan is perfectly closed by the lid
- How the hot air is blown through the sand for 5 mins
- After 5 mins the sand is reweighted

Merits

- The method is fast

Demerits

- Cost is high
- Moisture teller based on chemical reaction;
- In that machine the weighed quantities of both sand
- Put in the pan without mesh screen
- Calcium carbides react only with moisture present in sand.
- It produces acetylene gas
- The moisture content is determined

1.5.3. Grain Fitness Test

- The test is carried out a dry
- The order of decreasing the sleeve sizes from top to bottom
- So the top most sleeves is the coarsest
- The pan is placed under the bottom
- To test the fitness, the sand to be tested
- To vibrate for 15 mins
- Each % is multiplied by a factor

1.5.4. Clay Content Test

- A quantity of sand sample distilled water and 1% sodium hydroxide are put in mixing device
- The mixer is timed about 5 min
- The dirty water is removed from the top

- Distilled water are again needed with sand
- The water is drained completed

1.5.5. Permeability Test

- Its defend as the tendency of sand which allows the escape gases through it
- When it's originated in the moulding sand due to pouring of hot molten metal
- A weighted quality of sand is taken
- Both clay content are added
- Sand specimen prepared by a specimen rammer
- It consist of crank lower cam weight ram tube
- Then it's placed under the specimen pusher
- Then the pressure is applied by the weight through a specimen pusher.
- The pressure is measured by a dial gauge provided on top of the rammer
- It's done by operating the crank lower
- The cam will rotate by moving the rammer
- After applying sudden impact gets compressed form

1.6. Methods

- Standard method
- Rabid shop method

1.6.1. Standard Method

- The test consist of water seal pressure monometer, ball jet
- 2000 cc of air is filled in the jet
- The air is allowed through the specimen
- Now the level of mercury monometer changes
- At the same point the air entering the specimen becomes equal to the air passed through it
- The time is required using stop watch

1.6.2. Strength Test

- This test is mainly carried out to measure the holding power
- Compressive strength test
- Shear strength test
- Tensile strength test

- Bending strength test
- It's very important used universal testing machine (UTM)
- The test specimen is prepared in sand specimen
- Indicates of high and low strength
- The hand wheel is rotated to hydraulic mechanism
- The hydraulic pressure applied and the specimen
- The indicator are mounted on the tester
- One is testing low strength sand and the other one of high strength sand

Diagram

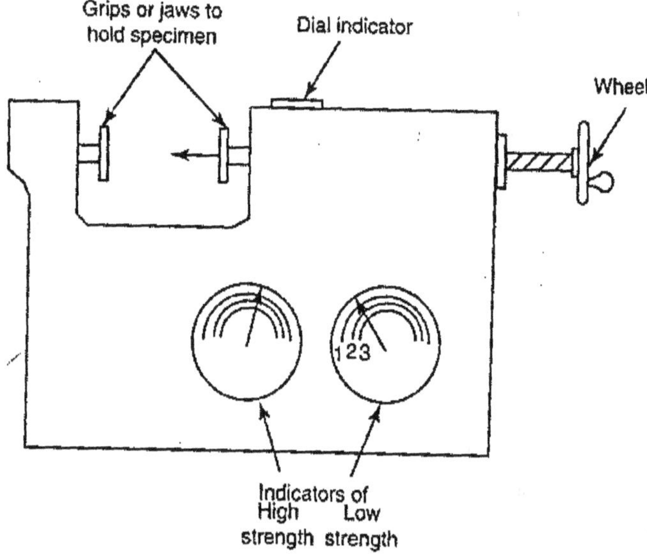

Figure 1.4: Strength Tester

1.7.   Deformation of Toughness Test

- It defined as the plasticity of sand that can be tested by reducing length of specimen applied
- High deformation indicates the better capacity
- Toughness refers the ability of sand to with stand handling pattern is drawn.

1.7.1. Hot Strength Test

- The cylindrical specimen with double end rammer
- The specimen is placed in a dilatometer
- For testing the specimen is heated from 500 to 2500 f compressive force is applied over the specimen

1.7.2. Refractoriness Test

- The heated resistance curved platinum strip is against capacity of moulding sand
- The specimen is placed under the fire about 2 hrs and 1550°c if the change in dimension is less than.

1.7.3. Mould Hardness Test

- It indicates the ramming density of the actual sand mould
- The test specimen is prepared
- Its placed on the table the indentation is formed
- It's formed on the specimen by rotating the manual loading device
- The force is applied on the specimen
- Then the specimen is taken out diameter of indentation is measured
- The mould hardness number is determined
- The scale provided on the tester itself will indicate the hardness number

Diagram

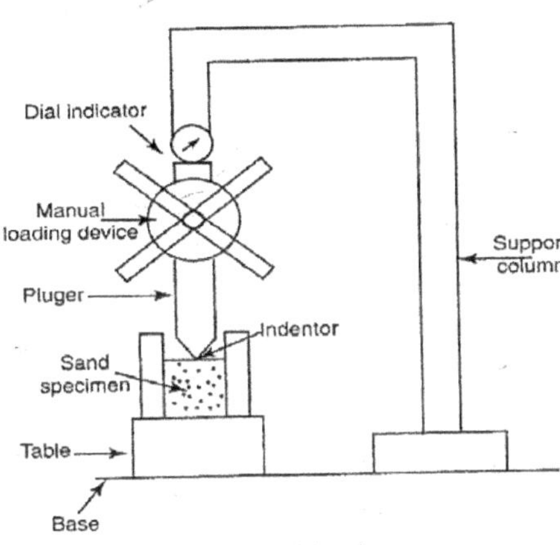

Figure 1.5: Mould Hardness Test

1.8. Cores

- A core is a body made of sand it's used makes a cavity
- It's also used to make recess, internal cavities
- It's an projection of a pattern it forms a sheet on the mould

1.8.1. Types of Cores

a) According to State of Core

- Green sand core
- Dry sand core

b) According to Position of Core

- Horizontal
- Vertical
- Balanced
- Hanging
- Drop

Tabular Column

- Core making methods
- It mainly made with machines
- It made by hand in core box

1.9. Types of Core Making Methods

- Hand core making
- Hot core box
- Synthetic resins based cold curing method
- Cold curing CO_2 process

1.9.1. Hand Core Making Method

- Introduction
- Diagram
- Stages
- Working
- Merits

Introduction

- Cores are made by hand in core boxes

Diagram

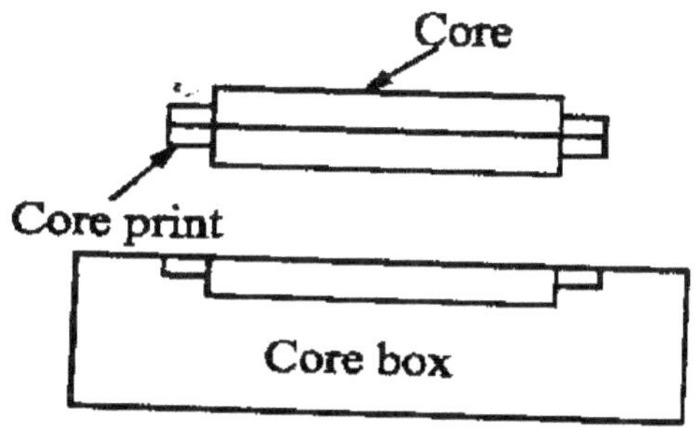

Figure 1.6: Cores

Stages

- Core sand preparation
- Moulding a green sand core
- Baking or curing
- Finishing
- Coating

Working

- The core materials of core sand already prepared
- It's thoroughly mixed with water
- A core is moulded by hand
- Its ram well after ramming core box is separated
- This known as green sand core
- The cores are heated in the core baking ovens at temperature 200^0c to 350^0c
- A surface coating is given to the finished cores
- Coating is applied by spraying
- That is also called dressing
- Hot core box method
- It provide strength and hardness to the cores

- The baking process can be eliminated by using quick set core sands
- Its hardening at 230-250⁰c in short time 2 to 3 mins
- The metal core box is heated by gas
- After holding time the box is opened. Withdrawn the core in the core box

Merits

- No separate oven is needed
- Good gas porosity
- Smooth and accurate
- Synthetic resin based cold curing method;
- These sands do not require further heat treatments
- Catalyst are added to speed up the hardening of binders
- The mixed sand is powered and get required compressive
- The core is then removed from the box
- The strength is less than the strength of 800 to 1200 kpa.
- It's used for production of cores

1.10. Moulding Process

- It's the process of making a mould cavity by packing the prepared moulding sand around the pattern

1.10.1. Types of Moulds

- Green sand mould
- Dry sand mould
- Loam mould

1.10.2. Green Sand Mould

- Introduction
- Diagram
- Working
- Merits
- Demerits

Introduction

- It's mostly used in the moulds
- It's poured immediately after the mould is prepared

Diagram

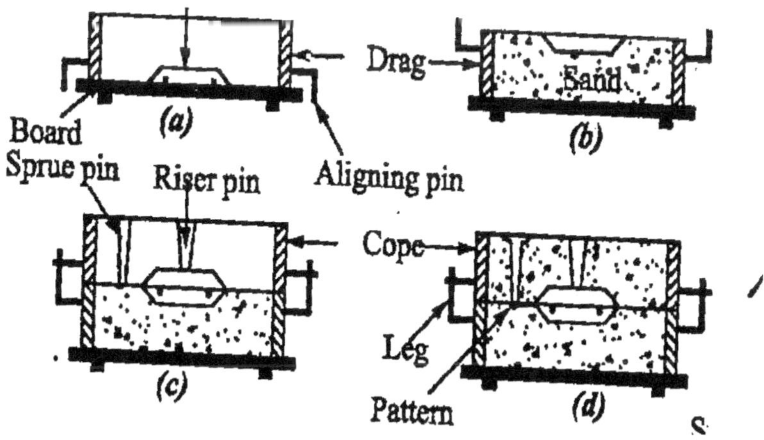

Figure 1.7: Green Sand Mould

Working

It making small size castings

- Non-ferrous metals two piece pattern is used
- 20 mm layer of facing sand it's placed around the pattern
- Excess sand is removed by strike off bar
- The top surface is made as smooth by towed
- The drag is tilted upside down
- It's placed in correct position
- Finally the drag are assembled
- The mould is ready for pouring

Merits

- Less expensive
- It can use in all metals

Demerits

- Strength is low
- Surface finish is low

1.11. Moulding Machines

- It's used for mass production
- Since hand moulding is slow process
- It reduces the cost and increase the quality of the mould

Operations

- Ramming the moulding sand
- Rapping the pattern for easy removal
- Removing the pattern from the sand

1.11.1. Types of Moulding Machines

- Jolting machines
- Squeezing machines
- Sand slingers
- Tabular column

1.12. Melting Furnace

- Blast furnace-melting iron to pick iron
- Cupola furnace
- Open hearth furnace-for steel
- Pot furnace
- Electric furnace
  - Direct arc furnace
  - Indirect arc furnace
- Conduction furnace

1.12.1. Blast Furnace

- Introduction
- Diagram
- Working
- Merits

Introduction

- It's used for melting metal ore, usually iron ore
- The very high temperature developed inside the furnace

- Its height is about 30 m and interior diameter 8 m

Diagram

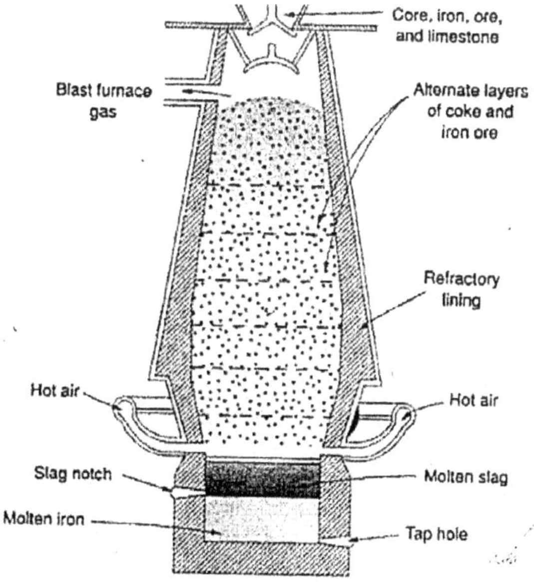

Figure 1.8: Blast Furnace

Working

- In this furnace unwanted silicon and other impurities are lighter than the pig iron
- The build-up the tall chimney like structure
- Coke lime stone powdered into the top
- The blast allows the combustion of the furnace
- The air blown into the furnace reacts with carbon in the fuel to produce carbon monoxide
- Which then mixes with the iron oxide reacting chemically
- Furnace is heated about 1500^0c
- The pig iron produced by blast furnace
- It is used to make cast iron goods
- When it escapes from the top of the furnace, further improving efficiency
- 60000 tonnes of iron per week produced in the furnace

Merits

- It is exist method
- Operations are simple

1.12.2. Cupola Furnace

- Introduction
- Diagram
- Construction
- Working
- Application
- Merits

Introduction

- It's used for melting iron

Diagram

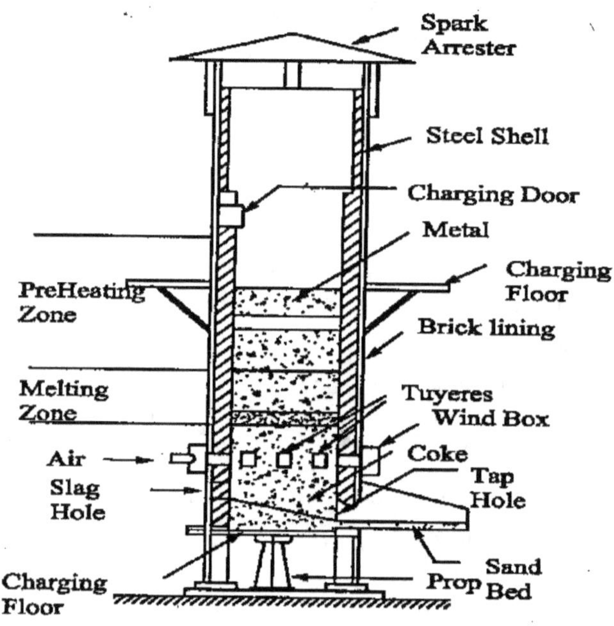

Figure 1.9: Cupola Furnace

Construction

- It's a vertical, cylindrical shell made of 10 mm thick
- It's lined with bricks inside
- Two bottom doors close the bottom of cupola
- A plug made of clay closes the top hole
- An opening called one meter above the bottom
- For charging the metal of fuel into the furnace

Working

- A slag and waste from previous melting are cleaned
- Brocken bricks are required
- A sand bed with towards top hole is prepared
- Wooden places are placed at the bottom
- The fire is started when the coke is charged at several positions
- The coke added to tuber level
- Finally the charging is done through the changing door.
- Pig iron are charged into the furnace
- The cupola is fully charged
- Then the molten metal can directly be poured into metals
- At the end the cupola is shut off by stopping the air black
- Wastes are dropped down.

Application

- It's used to melt cost iron

Merits

- Simple design
- It requires less area

1.13. Special Casting Process

- Shell Moulding Methods
- Investment Casting

1.13.1. Shell Moulding Methods

- The shell moulding casting is semi precise method
- The process involves the use of match plate pattern

Drawing

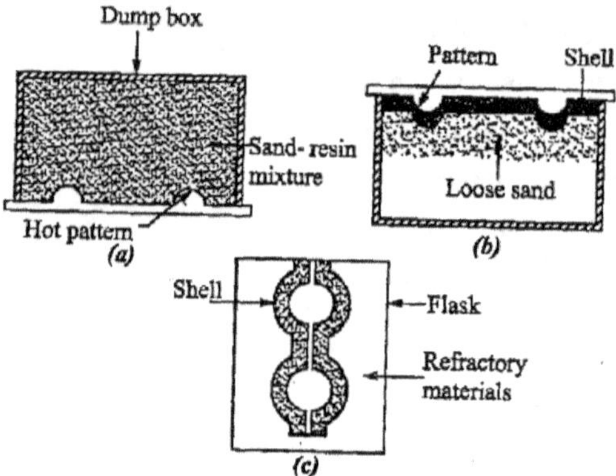

Figure 1.10

Construction and Working

- Patterns are machined from copper alloys
- Then it's attached to the metal match plate
- The mould material contains 5 to 10%
- It's noted that there is no water used
- The pattern is heated
- The heated pattern moulds and hardens to resin
- After a specified time 20 to 30 sec the pattern sand are inverted
- The liquid metal to be poured into the mould
- After cooling the shells are broken

Uses

- Making the braking drums
- Cams, piston rings are made

Merits

- Good surface finish
- Complex parts can be made

Demerits

- The cost is more

1.13.2. *Investment Casting*

- Introduction
- Diagram
- Construction
- Working
- Uses
- Merits
- Demerits

Introduction

- It have very smooth surface
- It's called as precision investment casting
- It means the layer material covered to make a mould
- Lost wax method
- Pattern are made of wax is melted

Diagram

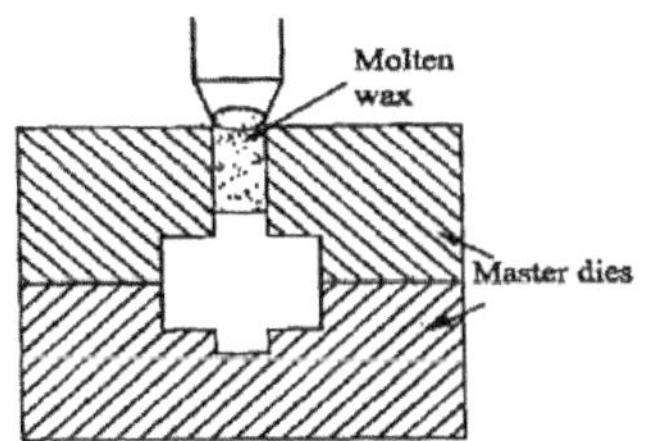

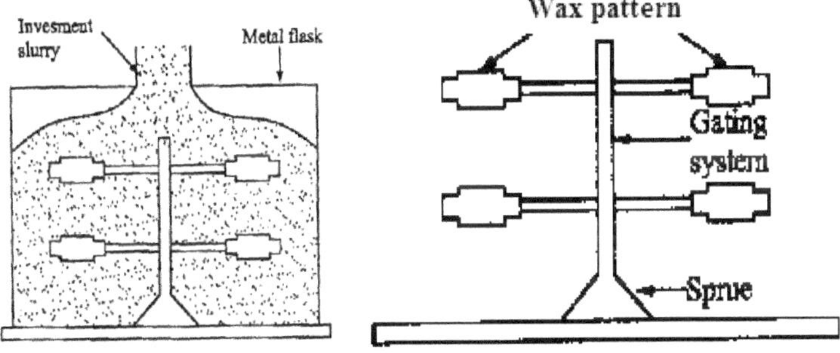

Figure 1.11: Investment Casting

Construction

The master pattern which is equal to the part to be cast is made by metal that can be easily machined

- Such as brass, fusible alloy

- The dimension of the pattern is larger than the size of part to be made

- It makes the pattern is very costly

- It makes a die out of a soft material

- The wax is injected

- The expandable pattern is rinsed in alcohol grease and dirt

Working

- After dying, a thin coating of primary investment slurry
- It contains the silica, water.
- The slurry contains being in direct with the surface of the wax pattern
- Sometimes a number of expandable patterns are assembled as a tree
- Fine grain silica sand is sprinkled over the wet surface
- Thus the produced coating on the pattern after drying
- It contains type of flask secured to the base by molten wax

Uses

- It produces nozzles, blades
- It used in costume jeweller

Merits

- Surface finish is very good
- High accuracy

Demerits

- The process is more expansive

1.14. Pressure Die Casting

Introduction

- In the die casting process, the mould used for making a casting is permanent called a die.

- Molten metal forced into the mould cavity at high pressure of low melting temperature material.

Types of Pressure Die Casting

- Hot chamber die casting.
- Cold chamber die casting.

1.14.1. Hot Chamber Die Casting

In hot chamber die casting, the melting furnace is an integral part of the mould.

- There is plunger at the top of the aeroseneck vessel.
- When the plunger is in the upward position.
- The molten metal flows into the vessel through a part provided on the side wall.
- The plunger is operated by hydraulic systems.
- The operating principle of hydraulic plunger is 15 MN/m² hot chamber die casting most suitable for zinc, lead.

Diagram

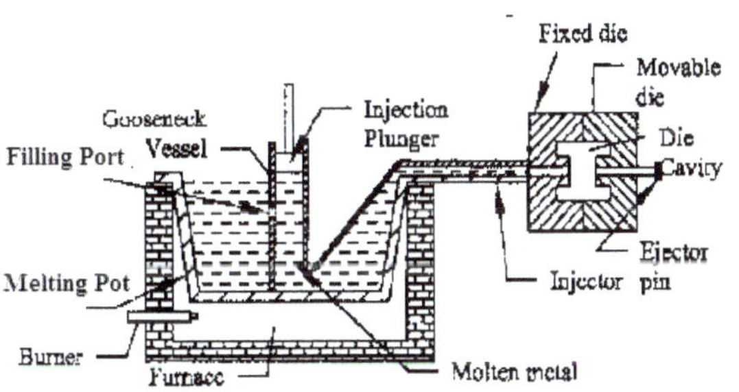

Figure 1.12: Hot Chamber Die Casting

1.14.2. Cold Chamber Die Casting

- In cold chamber die casting, the metal melting unit is not an integral part of the machine.
- The metal is melted in a separate furnace & brought to the machine for powering.
- Cold chamber of cylindrical shape with a hydraulic plunger.
- A measured quantity of molten metal is powered into the injection cylinder.
- Then the plunger moves to the right and forces the molten metal in to the die cavity.

Diagram

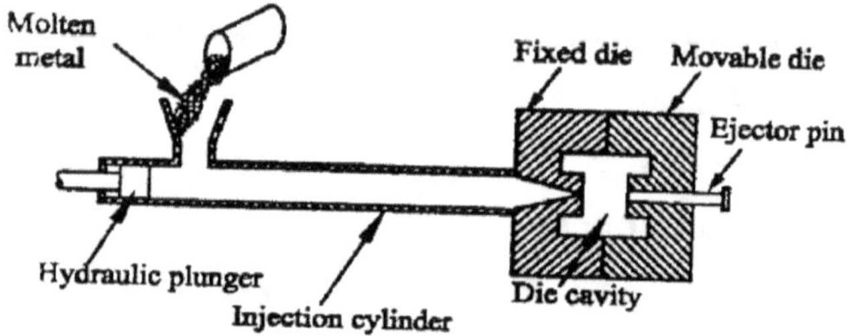

Figure 1.13: Cold Chamber Die Casting

Uses

Automobile parts like:

- Fuel pump
- Carburettor body.
- Telephones, television sets, speakers.

Merits

- Rate of production is high.
- Casting with very good surface finish.
- It is used for long time.

Demerits

- Equipment cost is high.
- Mass production.
- Nonferrous metals can be cost.

1.15. Centrifugal Casting

- Introduction
- Diagram
- Parts
- Working
- Uses
- Merits
- Demerits

Introduction

- Centrifugal casting is primarily used for making hollow castings, such as pipe without using core.
- In this process, a metal mould is made to rotate.
- The rotating mould is mounted on the trolley.

Diagram

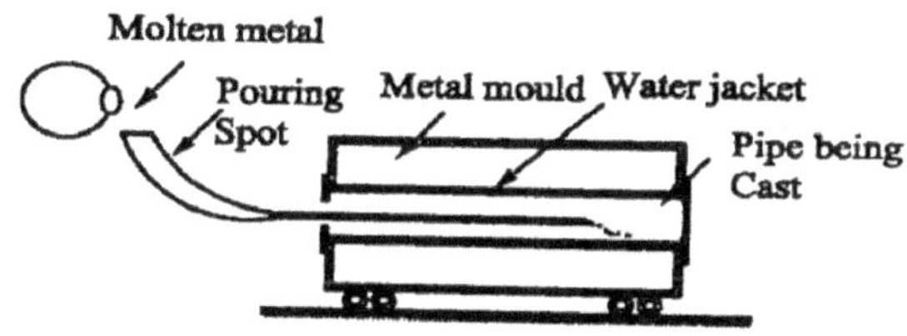

Figure 1.14: Centrifugal Casting

Parts

- Metal mould.
- Water jacket.
- Pipe being cast.
- Pouring spot.
- Molten metal.

Working

- The trolley moves over rails.
- The end of the mould is closed by end cores to prevent the flow of metal.
- The metal is powered is to the mould through a long spout.
- The mould is rotated by electric motor or mechanical means as well as moves axially on the rails.
- The outside of the mould is water cooled.
- The centrifugal casting method to produce the symmetrical and cylindrical objects.

Uses

- Components such as water pipes, gear
- Fly wheel, piston ring, brake drums
- Gun barrels.

Merits

- Rate of production is high.
- Wire is not required.
- Pattern, runner, riser are not required.

Demerits

It is suitable only the symmetrical shaped casting or circular object.

- Cost is high.

1.16. Cold Curing CO_2 Process

- Introduction
- Diagram
- Explanation & Working
- Advantage
- Disadvantage
- Uses

Introduction

- This process is used to make good casting.

Diagram

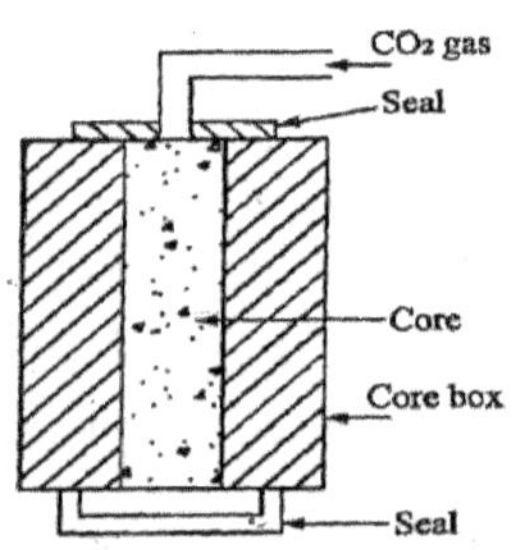

Figure 1.15: CO_2 Process

Explanation & Working

- It is used to make the good quality casting.
- Core material consisting clean dry sand mixed with solution.
- The mixing done is Muller.

- Sand mixture rammed in core box.

- Core box based 30 sec, pressure of 130-140.

- It forms sodium carbonate & silica int.

- That used sand mixture cannot be recovered.

Advantage

- Braking is not required.

- Semi-skilled operator is required.

Disadvantage

- Cores are susceptible to moisture.

- Due to inorganic nature of bond.

Uses

- It is used for producing high amount of cores and iron.

1.17. Stir Casting

- Introduction
- Diagram
- Parts
- Process
- Advantage

Introduction

- Stir casting is a method of composite materials fabrication in which a dispersed phase is mixed with a molten metal by means of mechanical stirring.

- Stir casting is a liquid state method

Diagram

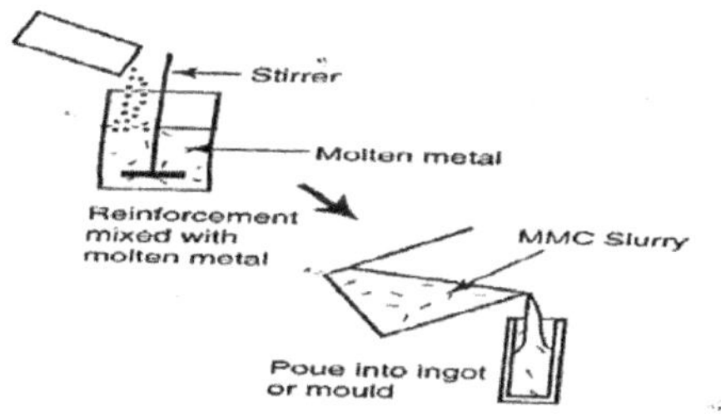

Figure 1 16: Stir Casting

Parts

- Stirrer.
- Molten metal.
- Mmc slurry.

Process

- Stir casting of mmcs involves producing a melt of the selected matrix material and obtaining the suitable dispersion through stirring.
- Solidification of the melt containing suspended partials to obtain the desired distribution of the dispersed phase in the case matrix.
- The melt changes the viscosity of melt and has implications for casting process.

Advantage

- Simplicity.
- Flexibility.
- High productivity.
- High volume production.
- Defects in sand casting;
- Design of casting and pattern
- Moulding and design of mould
- Malting and pouring
- Metal composition
- Gating and rising

Defects and Remedies

Defects name	Diagram	Descriptions	Causes	Remedies
Shrinkage		Depression of cost	Improper solidify	Proper solidify
Swell		Enlargement costing	Soft ram	Proper ram

Defects name	Diagram	Descriptions	Causes	Remedies
Scab		Erosion of portion	Quick pour Improper ram	Correct Ram
Blow Holes		Molten metal is poured and steam are formed	Improper pour Excess binder Hard ram	properly Proper pouring Control binder Ram properly
HARD SPOTS		It surface becomes very hard	Rabid cooling Pouring is 29improperly	It's not necessary give uniform
CRACKS		Corner of the casting	Due to sharp corners	Provide tuber
RUN OUT		Leakage of metal	Improper parting line	Provide proper parting line
HONEY COMBING		Number of small cavities present	soft ramming	Provide ram
SHAFT		miss matching of casting section	Loose dowels in pattern	Tight in dowels

Defects name	Diagram	Descriptions	Causes	Remedies
INFUSIONS		Foreign material in casting	Improves skimming	It proper skimming
COLD SHAFT		It completes filling of the mould	Too small gate	Provide correct gating system
FINS	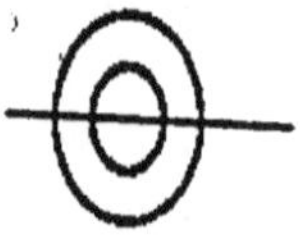	Project parting line	Pouring in correctly	Correct pouring.
DRESS		Lighter impurities appearing on the top of surface	Improper using of strainer	Use strainer properly
RAT TAIL		It's a long shallow, depression normally found in s thin casting	Improper compressed when expanding	Provide proper expansion
BLISTER		Scar covered by the thin layer of metal	Improper pouring temperature	Pour at a correct temperature

Unit 1

Metal Casting Processes

Part-A (2 Marks)

1. Name four types of commonly used patterns
2. What is the merit of CO_2 process?
3. State the essential properties of mounding sand
4. Give any two merits and demerits of investment casting process
5. Mention any two merits and demerits of die casting
6. List out any four defects in casting
7. What is meant by split pattern?
8. Define the term mould
9. What are the defects caused by low pouring temperature?
10. What is meant by match plate pattern making?
11. How will you calculate grain fineness number?
12. How patterns differ from casting?
13. What are the tests carried out to determine the quality of casting?
14. What are the functions of riser?
15. What are core prints?
16. What are the functions of gating and risering?
17. What is the composition of moulding sand?
18. What is the function of core?
19. Which process is called "Lost wax process"? Why?

Part-B (16 Marks)

1. i. Discuss the properties of moulding sand

   ii. What are the various moulding methods, explain them
2. i. Explain the working principle of investment casting

   ii. Discuss the casting defects and their inspection methods
3. i. What are the pattern making allowances and briefly explain them

   ii. Describe centrifugal casting process
4. i. Describe the shell moulding process

   ii. Explain the ceramic moulding process and state its merits and demerits

5. i. What are the factors which govern the selection of a proper material for pattern making?

ii. What are the specific advantages of match plate patterns? Explain how they are used for making mould

6. i. Classify the types of patterns and sketch any three of them

ii. What is core and explain how to make a core?

7. i. Explain the construction and operation of Cupola furnace with diagram

ii. Write a short note on "Chills"

8. i. Describe various materials used for making patterns. What are its merits and demerits?

ii. What are the basic requirements of core sand? How does it differ from the moulding sand?

9. i. What are the different types of furnace used in foundry? Describe in detail with neat sketches any one of them

ii. Describe the steps involved in the preparation of green sand mould with cope and drag pattern

10. i. Briefly explain cold-chamber die casting process with a neat sketch

ii. What are the advantages of centrifugal casting?

UNIT 2

WELDING

2.1. Classification of Welding

- Oxy-acetylene welding
- Oxy-hydrogen welding
- Air-hydrogen welding

2.1.1. Oxy-Acetylene Welding

- It's one type of welding process
- In which the edges of the metals to be welded are melted by using gas flame
- No pressure is applied

Diagram

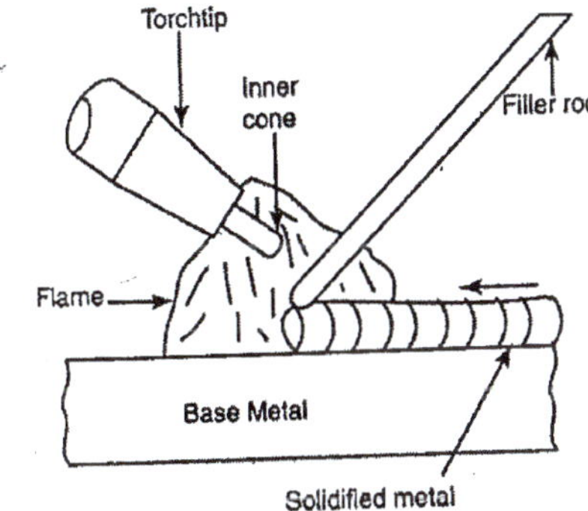

Figure 2.1: Oxy–acetylene Welding

Working

- The flame is produced at tip of a welding torch
- The heat is obtained by burning a mixture of oxygen
- The most form of gas welding is oxy-acetylene welding
- The flame will only melt the metal
- Weld is supplied by the fillet rod
- A flux is produced during welding to prevent oxidation
- Hottest region about 3200^0c

- The gases O_2 and c_2h_2 can be stored at high temperatures

There are Two Types

- High pressure system
- Low pressure system
- Oxygen is compressed to 120 atm gauge
- The acetylene is dissolve from 25 litres of acetylene
- The pressure of acetylene at the welding torch ions form 0.66 to 1 bar
- In a low pressure system acetylene is produced at the place of welding by interaction of calcium carbide and water in acetylene generator the chemical reaction
    - h.p system 0.1 to 3.5 bar
    - L.p system 0.5 to 3.5 bar

2.1.2. *Air Hydrogen Welding*

- It's similar to that of oxy-acetylene welding process
- Air is used instead of air
- The air taken from the atmosphere is compressed in a compressor
- The type of welding is limited use
- Its successfully used in lead welding
- Many low melting temperature metals and alloy
- This process is similar to Oxy-acetylene welding process.
- In this process special regulator is used in metering the hydrogen gas. Lead and magnesium but it is not in use today because more versatile and faster welding process.

2.1.3. *Gas Cylinder*

- For gas welding, a head of oxygen and acetylene are
- Used the standard colour for oxygen cylinder is black.
- The standard colour for acetylene cylinder is maroon.

2.1.4. *Pressure Regulator*

- Each cylinder is fitted with pressure regulator.
- The working pressure of oxygen is between 0.7 and 2. kg/cm^2. The working pressure of acetylene is between 0.007 and 1.03 kg/cm^2 depending upon the thickness of the work pieces to be welded.

2.1.5. Pressure Gauges

- There are four pressure gases provided in which two are placed on the oxygen cylinder regulators and two on acetylene cylinder regulators. The other one is for showing the working pressure for welding.

2.1.6. Hoses

- The regulator of each cylinder is connected to the torch through two long hoses. Oxygen cylinder is connected with black colour hose where as acetylene cylinder is connected with ted colour.

Welding Torch

- Two gases are entering the torch through the hose in separate passage.
- A flame will be produced at the tip of the torch called nozzle

2.2. Gas Welding Technique

- The speed and quality of the welding can be improved by the proper section of torch size
- There are two techniques are commonly used

2.2.1. Left Ward Welding

- The torch flames moves from right to left
- The torch is held on the right hand and the welding rod is held on left hand
- Its angle between 60 to 70
- The torch is given for very slightly sideways movement
- This technique is suitable for m.s plates up to 5 mm thickness.

Diagram

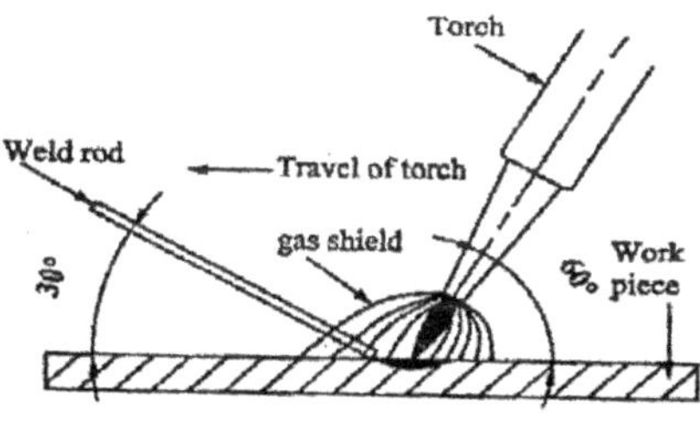

Figure 2.2: Left Ward Welding

2.2.2. *Right Ward Welding*

- The torch flame moved from left to right
- There is no sideways movement
- The torch is angle held between 40 to 50
- It provides a better shielding against atmospheric reaction

Merits of Gas Welding

- Flame can be easily controlled
- The flame can be used welding and cutting
- All types of metal can be welded
- The cost of equipment is less

Limitations

- It's not suitable for joining thick plates
- It's a slow process
- Comparison of A.C Machine and D.C machine welding machines.

Tabulation 2.1

S.No	A.C Machines	D.C Machines
1.	Efficiency is more	Efficiency is less
2.	Power consumption is less	Power consumption is more
3.	Cost of equipment is less	Cost of equipment is more
4.	Its noise less operation	It's very noiseless operation
5.	Any terminal can be connected	Positive terminal connected to work
6.	Voltage is higher ,hence it's not safe	Voltage its low hence its provides operation
7.	It's not suitable for welding nonferrous metals	It's very much suitable

2.3. Gas Tungsten Arc Welding

- Gas tungsten arc welding is called TIG Welding
- The electric arc produced between non consumable tungsten electrode and the work piece

Diagram

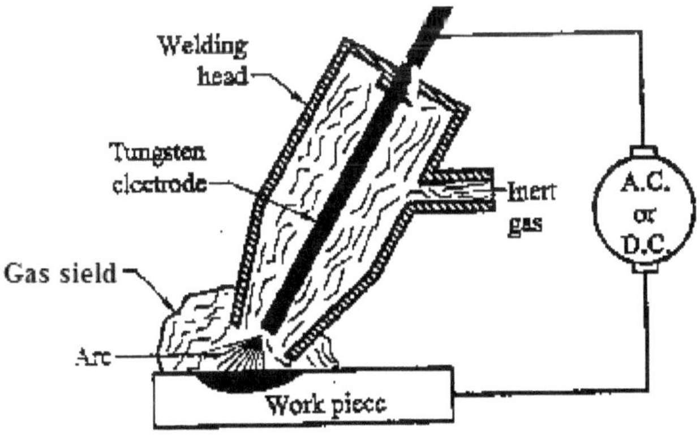

Figure 2.3: Gas Tungsten Arc Welding

Working

- The supplying the electric power between the electrode and the work piece
- The inert gases from the cylinder passes through the nozzle of the welding head around the electrode
- It protects the weld from atmospheric efforts
- Fillet material may not be used
- When fillet material is used manually into the weld pool
- It has high melting point
- process is used for welding steel, cast iron stainless steel
- It's also used for combining the dissimilar metal

Merits

- No flux is required
- The welding speed is high
- It can be used both ferrous and non-ferrous metals

2.4. Gas Metal Arc Welding

Introduction

- The process is also called as material inert gas welding

Diagram

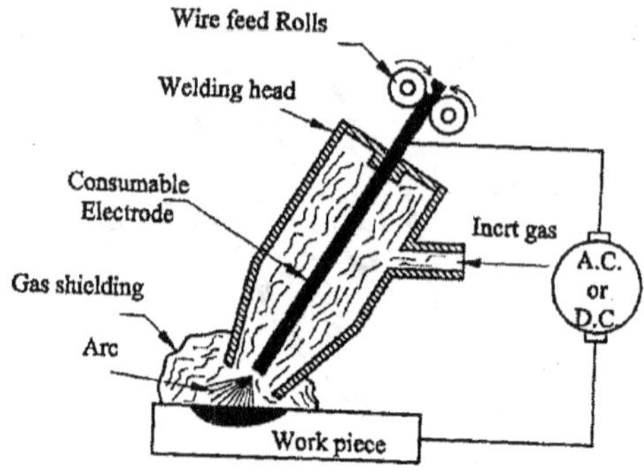

Figure 2.4: Gas Metal Arc Welding

Working

- The electric arc produced between metal electrode and work piece
- During welding the arc and welding zone are surrounded by inert gas
- The surrounded air protects the weld from atmosphere
- The electric is fed continuously through welding head because during welding the electrode is melted
- A.C generator is used for MIG welding
- The current ranges from 100 to 400 A

Merits

- No flux is required
- High welding speed is obtained
- The process is cheaper

2.5. Submerged Arc Welding

- Introduction
- Diagram
- Working
- Merits
- Demerits

Flux Core

- When the flux is required on that time the flux is used in the form of wire wound on rotating drum is called flux core
- Its mainly used avoid oxidation

Introduction

- The complete welding setup is dipped in the flux powder known as submerged arc welding
- It is also names as hidden arc welding

Diagram

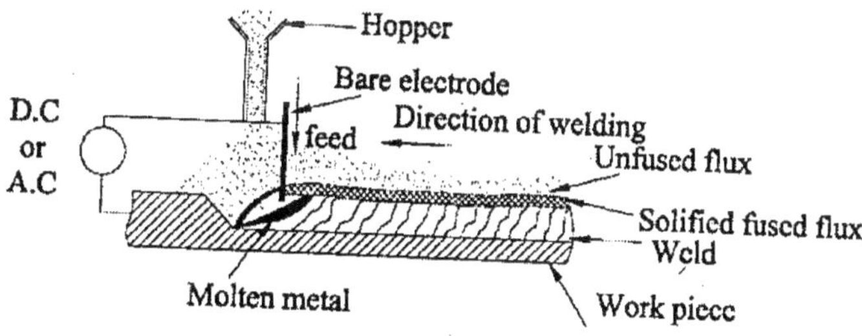

Figure 2.5: Submerged Arc Welding

Working

- The electric arc produced between electrode and the work piece
- The arc is completely sub merged ,hidden under the flux powder
- The arc is not visible out side
- Its supplied from a hopper
- When the arc is produced in the welding zone covered by the flux powder
- The flux powder is made of silica, metal oxides and other components
- It may also contain powder metal alloying elements
- The flux covers the arc and molten metal
- Some of the flux melts and forms the slag on the weld
- The un used flux is sucked a pipe
- Voltage used here is 25 to 40 v
- In this welding are used welding carbon steels

Merits

- It's a very fast method
- Deep penetration can be obtained
- Long joints can be easily welded

De-Merits

- It's not suitable for welding works which is inclined for vertical.

2.6.   Electro Slag Welding

- Introduction
- Diagram
- Working
- Merits
- Demerits
- Uses

Introduction

- Coalescence is formed by molten slag and molten metal pool remains shielded by the slag

Diagram

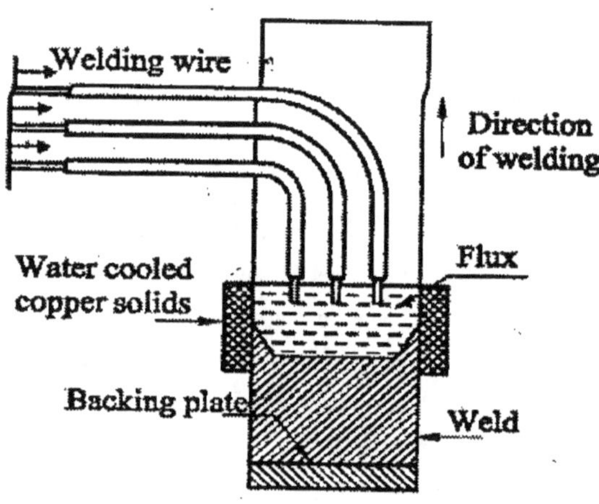

Figure 2.6: Electro Slag Welding

Working

- The electric arc struck between the electrode and work join by use of steel wood
- Welding flux is added and use of heat flux is added and further melted by use of heat from arc
- The action is stopped until the molten slag form
- The temperature ranges from 1600 to 1900^0c
- So this high amount of heat energy is enough for melting the work piece
- Thus the weld is formed

Merits

- Low stress formation
- Low distortion

De-Merits

- Hot cracking may occur
- The cost is high

Uses

- Heavy plates can be welded
- Forgings and casting are welded

2.7.   Resistance Welding

- The parts to be joined are heated by plastic state by their plastic
- There are two copper electrodes in a circuit
- The metal padres to be welded
- When the current passed through the electrode metal joints become very high
- Now the mechanical pressure is applied by complete weld
- $Q=i^2\,RT$
- Q=Heat

    I=current in amps

    R= Resistance of the assembly

    T=time of the current flow
- The power supply ranges from 6 to 18 kw
- It's used for mass production of welding sheet metal and wire and tubes

Types of Resistance Welding

- Spot welding
- Butt welding
- Seam welding
- Projection welding
- Stud welding
- Percussion butt welding

2.8.   Spot Welding

- Introduction
- Diagram
- Working

Introduction

- It's one type of electrical resistance process

Diagram

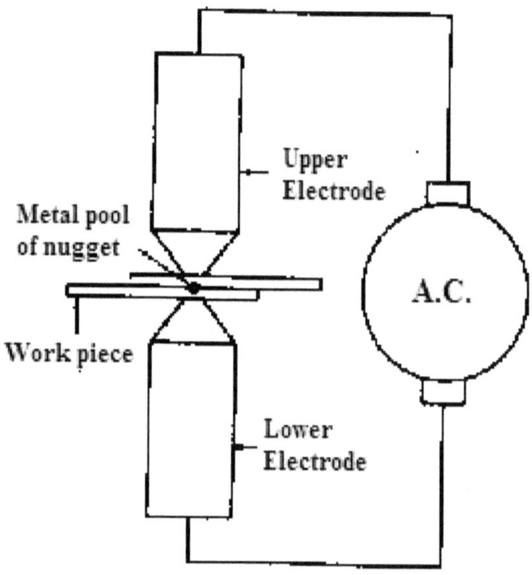

Figure 2.7: Spot Welding

Working

- The metal places are assembled and placed between the two copper electrodes when current passed
- The parts are heated by resistance
- Then the electrodes are pressed against the metal pieces by mechanical pressure
- Electrode possess high electrical and thermal conductivity
- The electrode pressure can e range up to 2 km
- It can be done an metal strips up to 12 mm thick
- All alloys can be spot welded

2.9. Butt Welding

- Upset butt welding
- Flash butt welding

Upset Butt Welding

- The parts to be welded are climbed in copper jaws
- The jaws acts as electrode
- They may be some gab between the parts
- Solid contacts when the current flows through the point of contact form locality of high electrical resistance
- It's mainly used for bars wire tubes

2.10. Flash Butt Welding

Diagram

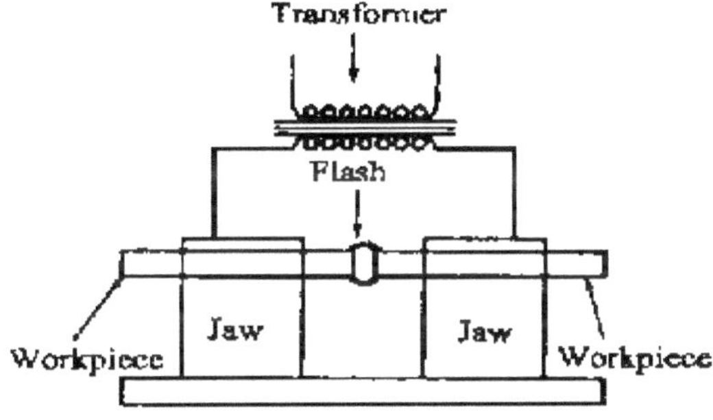

Figure 2.8: Flash Butt Welding

Working

- The parts to be welded are clamped in copper jaws of the welding machine
- They are connected to heavy current electrical supply
- The work pieces are brought together in a slight contact when the current flow through the work piece
- The fusing temperature and power is switched off
- Now the ends, applying mechanical force to complete the weld
- The projection is finished by grinding
- This process is suitable for welding steel
- Dissimilar metals with different melting temperature can be welded
- But welding is used an automobile construction of the body.

2.11. Seam Welding

Introduction

- The joint between two over lapping pieces of sheet metal
- The work pieces are placed between the rotating wheel electrode
- When electric current is passed through the electrodes
- High heat is produced on the work piece on wheels
- At the same time the pressure is applied completely the weld
- Finally speedup the welding process

Uses

- It also used for welding thin sheets

2.12. Percussion Welding

Diagram

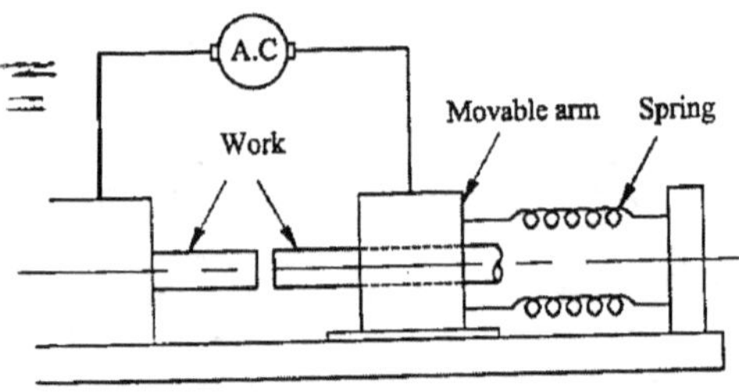

Figure 2.9: Percussion Welding

Working

- The parts are to be welded are climbed in copper jaws of the machine
- One clamp is fixed another one is movable
- Movable clamp is back against the pressure from a heavy spring
- The jaws act as electrode
- Heavy current is connected to the work pieces when the clamp is released.
- Causing an intense arc between two surface
- No upset occurs from the weld
- The method is limited from small areas

2.13. Projection Welding

Diagram

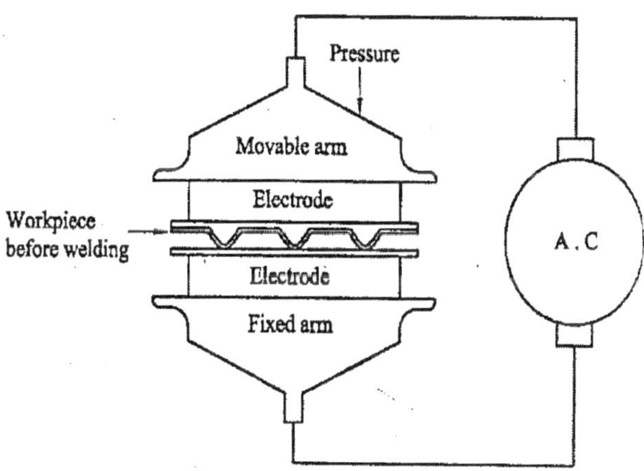

Figure 2.10: Projection Welding

Working

- The metal pieces are welded and placed between two metal arms
- One of the work pieces has projections on its surface
- The work pieces are clamped between the arms
- The surface of the projections must be cleared

- There should not be any scale on the surface
- It's used for joining thin sheet metals of thickness up to 3 mm
- It's used for mass production

2.14. Stud Welding

Diagram

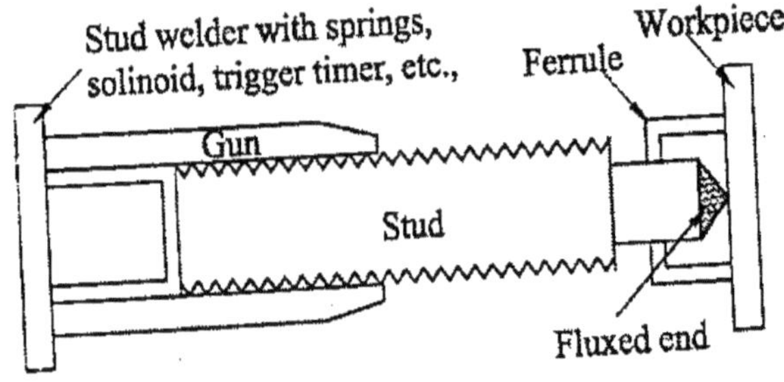

Figure 2.11: Stud Welding

Working

- The electric arc is produced between the stud and the work piece
- The arc melts the stud is forced on the metal surface
- A special welding gun is used for spring solenoid trigger timer, etc
- The front gun is held against the work piece
- When the gun is pressed, welding current flows between the end of the stud
- A molten pool is formed on the work piece
- Now the melted end of the stud is pressed on the molten pool of work on its surface
- The process is automatic and stud on bodies without drilling and tapping

2.15. Plasma Arc Welding

- Introduction
- Principle
- Diagram
- Working
- Uses
- Merits
- Demerits

Introduction

- The conventional method is not suitable for machining metals such as cast alloy, application is various industries in conventional methods causing increased machine cost

Principle

- Plasma is high temperature ionized gas
- It's a mixture of neutral atoms, charged atoms and free elements
- When this high temperature is passed through the orifice
- The proportion of gas is increased
- The plasma arc welding is formed

Diagram

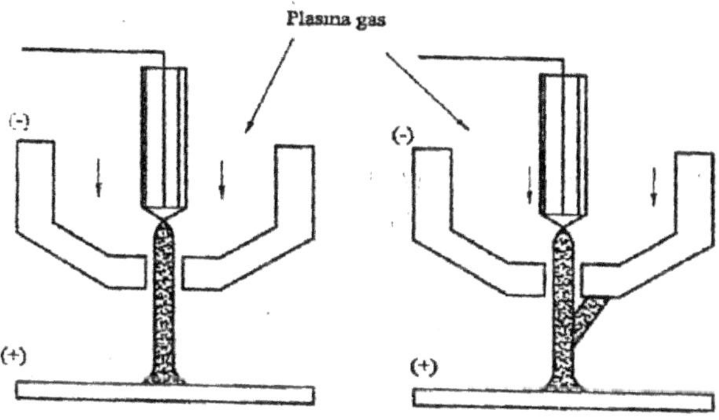

Figure 2.12: Plasma Arc Welding

Working

- The high heat content plasma gas is forced through the torch, the plasma cutting force imposes a swirl on the orifice gas flow
- The arc is initiated in the beginning by supplying electrical energy between nozzle and tungsten electrode
- This high amount of heat energy is used to weld the metal
  1. Transfer type
  2. Non transfer type

Transfer Type

- Tungsten electrode connected to the negative terminal
- Work piece connected to the positive terminal
- Electric arc maintains the electrode and the work piece
- It's difficult to initiate the arc first between the work piece and electrode

Non-Transferred Type

- Power is directly connected with the electrode and the torch of the nozzle
- The electrode carries the current
- That high velocity gas strewing towards the work piece
- The spots moves inside the wall and heat the incoming gas and outer layer remains pool
- This type plasma has low thermal efficiency
  Base metal welded by plasma arc welding are
    - Stainless steel
    - Titanium alloy
    - Copper alloy

Types of Joints

- Filler welds
- T-welds
- Grooves

Uses

- It's used in aerospace application
- It's used for high melting point
- It's used in welding nickel alloys

Merits

- Penetration is uniform
- Arc stability is good

De-Merits

- Gas consumption is high
- Its limited to high thickness application

2.16. Thermit Welding

- Introduction
- Diagram
- Uses

Introduction

- The welding parts by using liquid thermit steel around the portions to be welded is called as thermit welding
- It's a fusion welding process
- Neither arc is produced to heat parts
- For getting high temperature exothermic reaction is used

$$8Al + 3FeO_3 \quad = \quad 4Al_2O_3 + 9Fe$$

Diagram

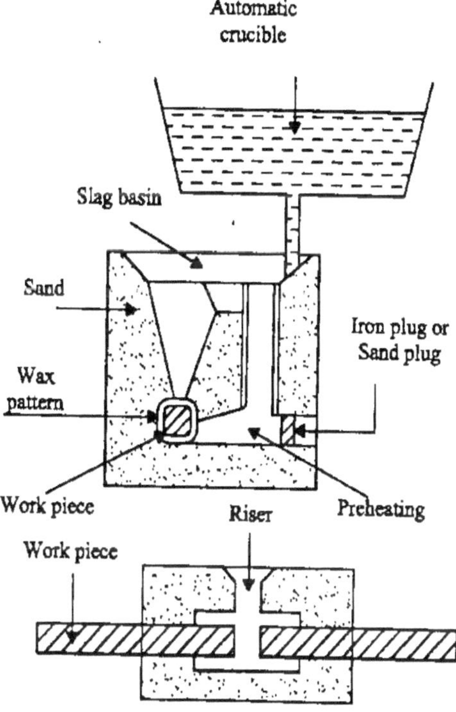

Figure 2.13: Thermit Welding

Classified Two Types

- Pressure welding process
- Non pressure welding process

Pressure Welding Process

- The parts to be welded are butted and enclosed in a mould
- The mould can be easily remove after weld the material
- First the heated iron slag is poured to the mould
- This will creating heating of parts
- Then the pressure is applied on join the work piece

Non Pressure Welding Process

- The parts to be welded are lined up in parallel and a groove is taken in the parts
- The wax pattern is removed
- Then the ram is around the wax pattern
- The sand is rammed around the wax pattern
- The mould is heated and wax is melted
- Finally the heated iron slag and aluminium are poured into the mould after solidification of liquid metal

Uses

- It's used in steel rolling mills
- Auto mobile parts are welded by this process

2.17. Electron Beam Welding

- Introduction
- Diagram
- Working
- Merits
- Demerits
- Uses

Introduction

- Beam of the electron is used for producing high temperatures and melting the work pieces welded

Diagram

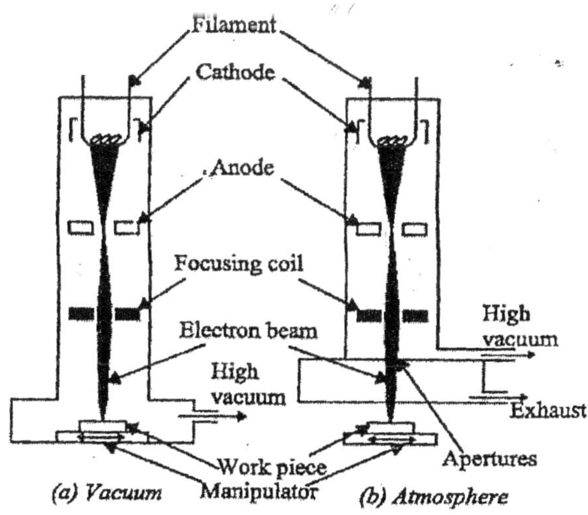

Figure 2.14: Electron Beam Welding

Working

- Tungsten filament is electrically heated in vacuum
- It will emits the electrons it carry a negative charge its passed through a anode hole
- Beam is focused by focusing lens
- Electron beam strikes the work piece
- Kinetic energy of this electron beam converted to heat energy
- This operation is carried in vacuum so it's possible to weld holes the beams are focused about 0.25 to 1 mm
- The variables which are controlled

  1. Voltage
  2. Speed
  3. Distance between beam guns to work piece

Merits

- High quality weld is produced
- Deep welding is possible
- Energy loss is very loss

De Merits

- The cost is high
- Skilled persons are required

Uses

- Dissimilar metals can be welded
- it's used in air crafts
- laser beam welding;

2.18. Laser Light Amplification by the Stimulated Emission of Radiation

- Introduction
- Diagram
- Working
- Merits
- Demerits
- Uses

Introduction

- light energy is converted into heat energy
- Here light energy produced from the laser source such as ruby rod.

Diagram

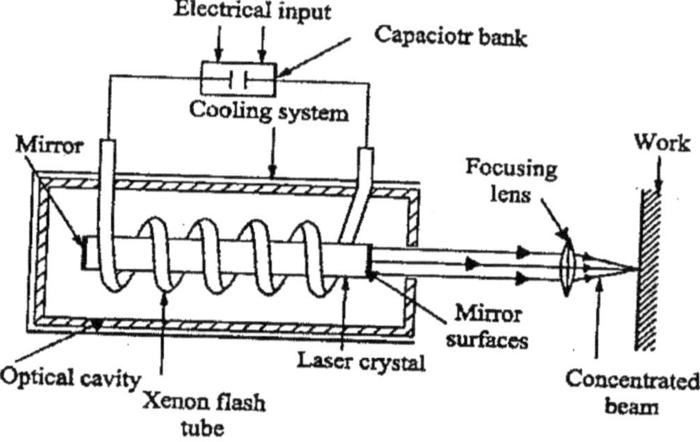

Figure 2.15: Laser

Working

- The electrical discharge from the capacitors the flash tube converts current energy to light flashes
- When ruby rod is exposed to the intense light flash the chromium atoms of the crystal pumped to a high energy level
- The laser light is not only intense but also can be readily focus without intensity
- When this light energy is impacted to the work
- To the work piece it will convert into heat energy

The Various Laser Forms are

- Liquid laser
- Gas laser
- Ruby laser
- Semi conduct laser

Merits

- It's used glass and plastics
- There is no need of electrodes and power
- Accuracy is greater

De Merits

- Welding process is low
- It's not suitable for large production

Uses

- It's very useful in electronic components in welding
- It can joint dissimilar metals

2.19. Friction Welding

- Introduction
- Diagram
- Working
- Merits
- Demerits
- Uses

Introduction

- It's a solid state
- Coalescence is formed by the heat is formed by the heat is obtained mechanical sliding motion between rubbing surface

Diagram

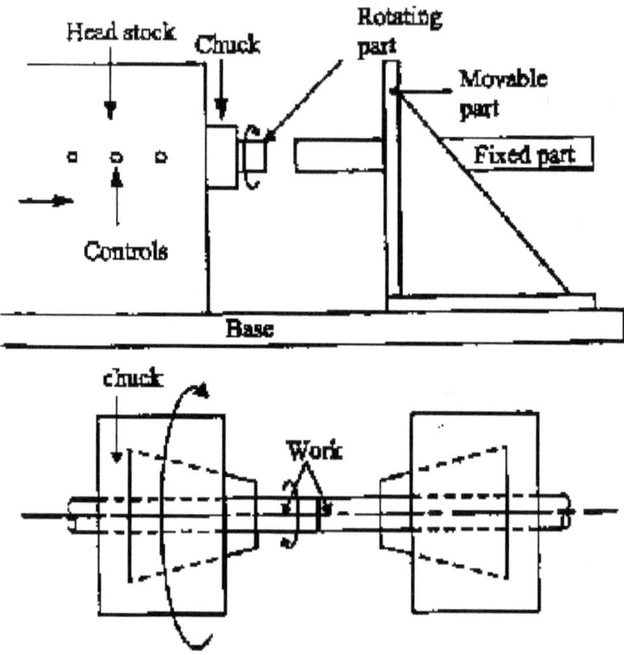

Figure 2.16: Friction Welding

Working

- They components to be welded are held under pressure
- One part is rotated and other part is stationary
- The movable clamp is moved and contacted with the rotating component
- The heat is produced between contact surface
- The heat is used to weld the components under pressure
- The parameters considered friction welds are
  1. Friction pressure

2. Speed

- Burn off

  The materials can be welded are listed

- Brass of bronze
- Nickel
- Titanium alloys

  The basic joints are made by friction welding:

- Bar-belt joint
- Bar-ball joint
- Tee-butt joint

Merits

- Power consumption is low
- The operation is easy

Demerits

- Heavy components are not used for weld
- There is possibility of heavy flash out

Uses

- It's used in super alloys
- It's used in refrigeration

2.20. Friction Stirs Welding

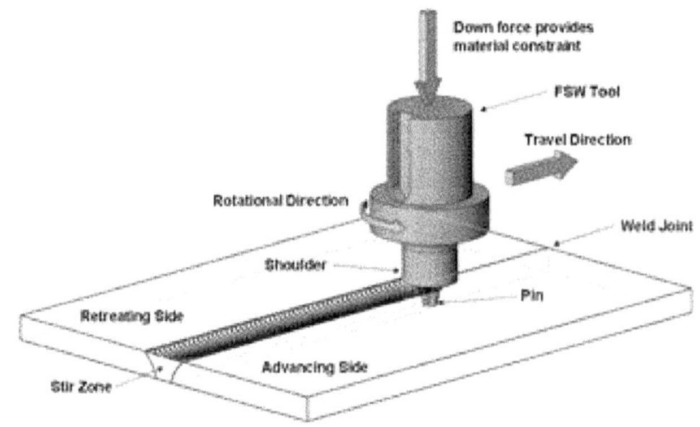

Figure 2.17

- It's a solid state welding process
- Rotating tool is fed along the joint line between two work pieces
- Work pieces
- During welding the heat is generated due to friction
- The metal is mechanically string to form weld stem

Working

- The rotating tool is consists of a cylindrical shoulder
- During welding the shoulder rubs against the top surfaces against top surfaces of the two parts.
- Developing friction heat mechanical mixing of the metal along the butt surface
- At the same time the proper has been designed to order perform the mixing perfectly
- The softening of metal takes place up to a highly plastic condition
- When the rod moves forward along the join ,rotating probe is forcing the metal around it
- So the shoulder helps to limit the practised metal flowing around the probe
- The main application are butt joints on large aluminium parts titanium as well as polymers are also used in few

Merits

- It permits less distortion
- It provides good weld appearance

De Merits

- Heavy duty clamping of the parts is required

2.21. Diffusion Welding

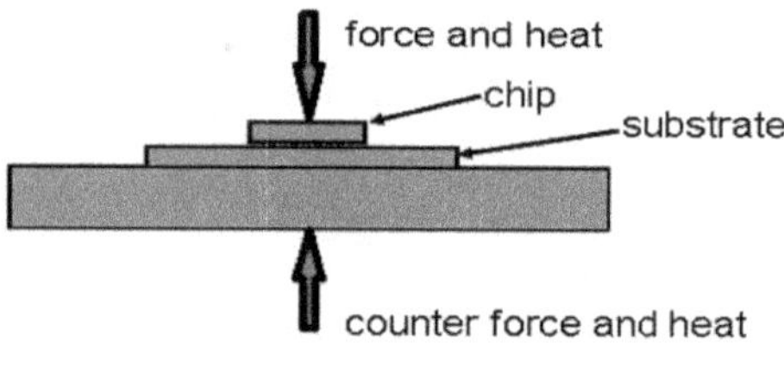

Figure 2.18

- It's a solid state joining process
- The process requires temperature about 0.5T
- In order to have a high diffusion rate between parts are joined
- The strength of the welding depends upon temperature time of contact clean less of metal
- The gold is then placed over copper and then weight is placed on top of it
- The pressure may be applied by dead weights
- The parts are usually heat by furnace
- The diffusion welding process slower process when compared other process welding process

Flame Cutting

- One of the easiest method of alloying steel
- The finished edges of finished parts are not as fine
- The process works on all steel products
- Flame cutting is used in the welding

Procedure

- Steel can catch fire and burn just such as coal wood.
- Pure oxygen is used to keep the firing
- This is a true chemical is not simply melted away
- Instead of leaving behind a material called slag which is basically iron oxide
- This method can be used up to 60 inches thick

Operation

- To start a cut, user brings a flame closer to the steel plate using a cutting torch
- It burns a mixture of fuel gas create a very hot pre heat flame
- The steel is heated to temperature 1500f
- When it's hot enough high pressure oxygen is turned on
- The metal will start burning at that instant velocity of oxygen flow
- It creates a hole in the metal all the way
- Now the cut has been started and the metal is hot enough
- The heated metal will burn and the slag will be flushed away
- More heat is generated and also the cutting process is obtained

Chemical Reaction

$$3Fe + 2O_2 = Fe_3O_4 + 267000 KCal$$

2.22. Brazing

Introduction

- Joining of two metal pieces by use filler material
- Liquid temperature is above 450^0c
- Base metal of the work pieces to be joined is not melted but filler makes the joint in metal
- The fillet of the metal in the form of liquid and the joint is made between two work pieces
- The fluxes are added to remove the oxides present in the filler metal then only base metal joint is cleaned
- Its commonly used fluxes are boric acid chlorides fluorides for brazing the ferrous materials

The Following Steps

- Surface should be cleaned before brazing the metal
- Flux is added to the liquid metal
- The filler is applied on the working pieces after it reaches the brazing temperature

The Flux Material Characteristics

- viscosity is low
- Melting temperature is low
- It has good surface tension

Merits

- Dismantling of joints is possible
- This process is quick and clean
- Complicated shapes can be made

Uses

- It's used in pipe fittings
- It's used in jewellery application

Limitations

- The cost is high
- Skilled labours are required

2.23. Brazing Methods

Torch Brazing

- The acetylene natural gas ,and butane are combined with air supply the required heat to melt the fillet rod
- It's not used for mass production

Furnace Brazing

- The heat is supplied by a gas
- The furnace can be either box type
- The performed shape of filler metals is placed on the parts to bar joined
- It's used for mass production

Induction Brazing

- The fillet metal is used in the performed shape
- The electric and magnetic resistances of the changing direction filed are produced in eddy current
- It is also used for mass production

Dip Brazing

- Chemical dip brazing
- Molten metal bath process
- Chemical dip brazing
- The parts with performed filler metal placed into a molten path
- It is mainly used for joining large parts

Molten Metal Path Process

- The assemble parts are first refluxed and then immersed into a molten bath or filler metal

Resistance Brazing

- The fillet metal is placed in the joint
- Laser brazing and electron beam brazing;
- It is a costly process

- It's used to join the metal
- It's used only precision work

2.24. Soldering

- Two dissimilar metals by means of filler metal is called soldering.
- Melting temperature is below 430^0c its being a low temperature 300^0c
- It is does not bring distortion

Solder Characteristics

- Its act as a good adhering film
- It welds the base meta

2.25. Classification

- Complete welding
- Partial welding
- Total non-welding

The Selection of Solder

- Melting and spreading characteristics
  - Resistance to corrosion
  - Material cost
- The selection of flux depends on
  - Cost
  - Flux type
  - Joint design
  - Temperature
- The various soldering techniques
  - Surface cleaning
  - Flux application
  - Surface tinning

Soldering Methods

Hard soldering

- Soft soldering
- Dip soldering
- Wave soldering

Hard Soldering

- The filler metal is silver it's also known as silver soldering
- This method is more expansive than plain brazing because the filler metal is more expansive
- It's used for joining electrical parts

Soft Soldering

- Most commonly used for soldering method
- Its copper rod with a tin tip which is used for flattering the soldering material
- The soldering iron can be heated by keeping in a furnace
- The fillet metal used in soft soldering is called soft solder

Dip Soldering

- Large amount of solder is melted in a closed tank
- The parts are to be cleaned
- If then dipped into the molten solder pool and lifted after completing the soldering process

Wave Soldering

- It is a different method in which the parts to be welded is not dipped into the solder tank
- But wave is generated in the tank
- So the metal solder comes up and joint is made
- The other methods of soldering:
  - Oven soldering
  - Resistance soldering
  - Induction soldering
  - Infrared soldering

2.26. Defects in Welding

- In complete fusion
- Cracks
- Porosity
- Under cut
- Distortion

- Slag inclusion
- Lamellar teaming
- Overlapping

In Complete Fusion

- This due to improper penetration of the joint
- The Para meters mainly affects the welding current
- If the current is very low, it is not sufficient to heat the metal all over the place

Diagram

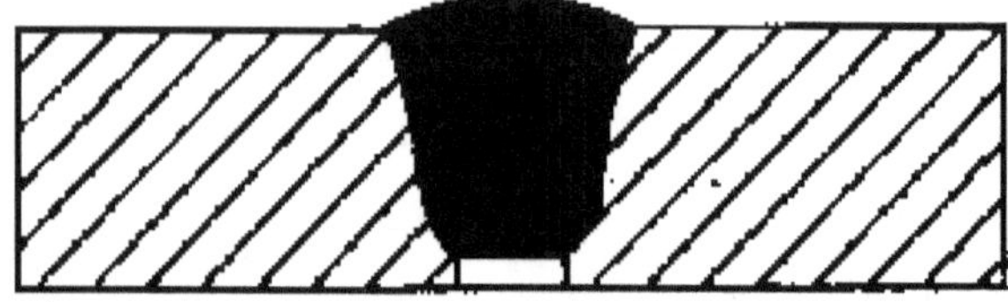

Figure 2.19: Incomplete Fusion

Cracks

- Hot cracking
- Cold cracking
- Hot cracking occurs at high temperature
- Cold cracking occurs at room temperature
- The main causes ,of crack formation
- Arc speed
- Ductility
- Solidification rate
- Temperature

Diagram

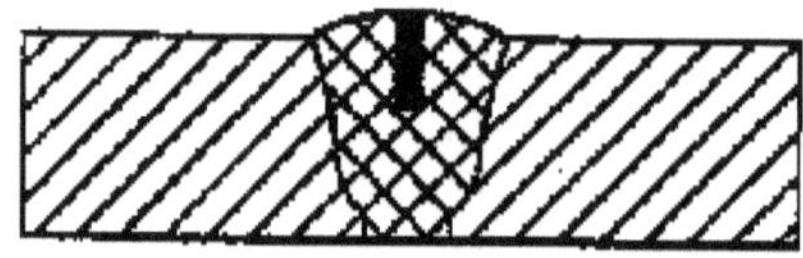

Figure 2.20: Cracks

Porosity

- Due to the presence of gases of the metal which are producing porosity
- The gases are oxygen ,nitrogen and hydrogen
- The parameters which are causing porosity are
  - Arc speed
  - Coating of the electrode

Diagram

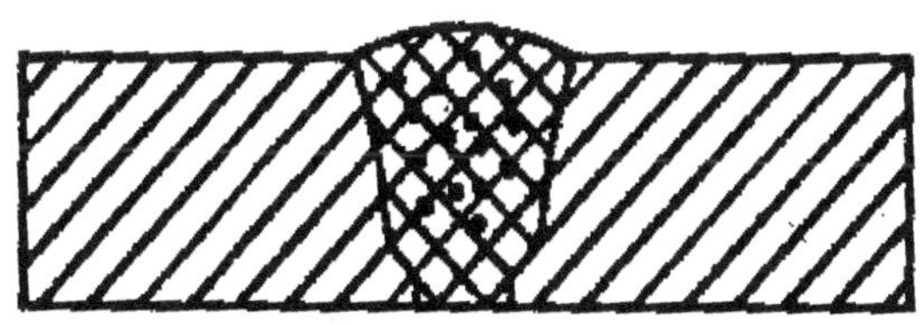

Figure 2.21: Porosity

Under Cut

- A groove gas formed in the parent metal along the sides of the weld

The Main Causes

- High current
- Arc length
- Electrode diameter

Distortion

- It is defined as the change in shape and difference between position of two plates during welding
- The base metal under the arc melts and already welded base metal start cooling

The Main Causes

- Arc speed
- Number of speed
- Stresses in plates

Diagram

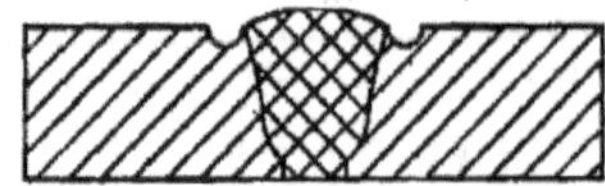

Figure 2.22: Distortion

Slag Inclinations

- During the solidifications of weld any foreign material present in the molten metal will not float
- It will be end rapped the metal

Diagram

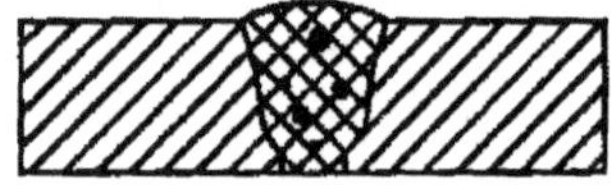

Figure 2.23: Slag Inclusions

Lamellar Drilling

- Presence of non-metallic inclusions
- Its formed during the non-metallic inclusions running parallel into the plate
- It is seen in larger structures

Diagram

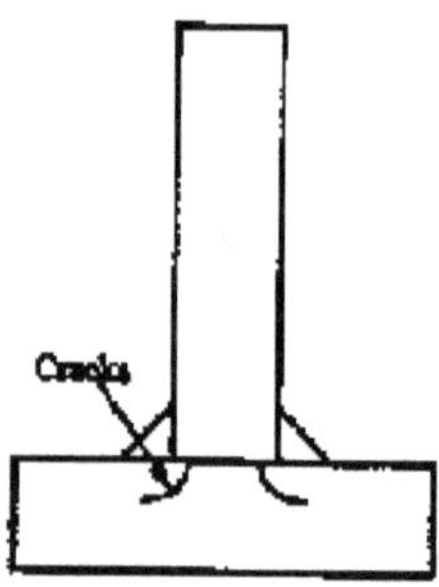

Figure 2.24: Lameller Drilling

Over Lapping

- molten metal flows over the parent metal and remains without fusing
- arc length
- arc speed
- joint type

UNIT 2

JOINING PROCESSES

PART-A (2 MARKS)

1. What is the principle of resistance welding?
2. What is the role of fluxes in welding? Or function of flux in welding?
3. List out any four arc welding equipment.
4. What is the principle of Thermit welding?
5. What are the different types of gas flames? How are they formed?
6. Differentiate soldering and brazing.
7. What is the chemical reaction occurs in Thermit welding?
8. What are the advantages of carbon arc welding?
9. Differentiate between oxy-acetylene and air-acetylene welding
10. What are the advantages of A.C. arc welding?
11. What is the principle cause of cracks in weld metals?
12. How do you specify an electrode?
13. What is the function of shielding gas in welding?
14. Why laser welding is used only for micro-welding applications?
15. Define resistance welding
16. What is flux? Why is it essential to use it in some welding situations?
17. What are the defects that are generally found in welding?
18. List any four applications of TIG welding process.
19. Is flux necessary in Brazing process? If yes why?
20. How will be avoided slag inclusions in welding?

PART-B (16 MARKS)

1. i. Distinguish between gas and arc welding

   ii. What are the advantages of welding?

   iii. Explain percussion welding

2. i. Describe Electro slag welding

   ii. Distinguish soldering and brazing

3. i. Explain spot welding

   ii. Explain submerged are welding

4. i. Explain the electron beam welding process with a neat sketch

   ii. Write a brief note on "Welding defects"

5. i. Sketch the three types of Oxy-acetylene flames and state their characteristics and applications.

    ii. Describe the electro-slag welding process with a neat sketch.

6. i. What is the principle of resistance welding and explain the seam welding?

    ii. Describe plasma arc welding

7. i. What are the different types of electrode? What are the functions of flux coating?

    ii. What is the principle of friction welding?

8. i. Describe metal inert Gas arc welding process with a neat sketch.

    ii. Briefly explain on butt welding process

9. i. Give a brief account of classification of welding processes?

    ii. Explain TIG welding process variables and enumerate its advantages

10. i. Describe shielded metal arc welding process with suitable diagram. What are its applications?

    ii. What is the difference between welding, brazing and soldering process?

UNIT 3

BULK DEFORMATION PROCESSES

3.1. Introduction

A product is produced by shaping the metal into the required shape and size. There are many methods to produce a product.

In a mechanical working method, "No machining process" is carried out it is used to achieve optimum mechanical properties in the metal.

By applying the force, the metal is plastically deformed into the metal reaches the "yield point".

Permanent (plastic) deformation of a material under tension, compression, shear or a combination of loads.

Bulk Deformation

Bulk Deformation–Significant change in surface area, thickness and cross section reduced, and overall geometry changed.

Ex

- Forging
- Rolling
- Extrusion
- Drawing

The metal forming processes are mainly classified into

- Hot working process
- Cold working process

3.2.   Hot Working Process

Synopsis

- Introduction of Hot working process
- Semi Hot working process
- Advantages of Hot working process
- Disadvantages of Hot working process

Introduction of Hot Working Process

Mechanical working of a metal above the recrystallization temperature but below the melting point is known as hot working.

It may also be defined as the plastic deformation of metals and alloy under the conditions of temperature and strain rate.

In this process, the metal is heated above the metal but below the melting point of metal (0.7 to 0.9 times of melting point temperature).

Semi Hot Working Process

In this method, the metals are deformed under the conditions of temperature and strain rate. The following factors should be considered for the selection of temperature and strain rate.

- Ductility
- Tolerance
- Yield strength

Advantages of Hot Working

- It is quick and economical process.
- Force requirement is less.
- This process is very suitable for all metals.

Disadvantages of Hot Working

- Surface finish may be poor.
- Tooling and handling cost are high.
- Sheets and wires cannot be produced.

3.3.   Types of Hot Working Process

The important hot working processes are

1. Hot forging
    a) Hammer forging
    b) Drop forging
    c) Upset forging
    d) Press forging
    e) Roll forging
2. Hot Rolling

3. Hot extrusions
4. Drawing
5. Swaging
6. Hot spinning

Hot Spinning

It consists of heating the metal to forging temperature and then it is formed into the desired shape on spinning lathe. It is used for thicker plates and sheets.

3.4. Cold Working

Synopsis

- Introduction of cold working process
- Materials used for cold working process
- Classification of cold working process
- Cold working process

Introduction of Cold Working Process

- Plastic deformation of a metal to the required shape being performed below the recrystallization temperature is known as cold working process.
- The recrystallization temperature is defined as the minimum temperature at which the complete recrystallization of a metal (cold worked)
- Takes place within a specified time.

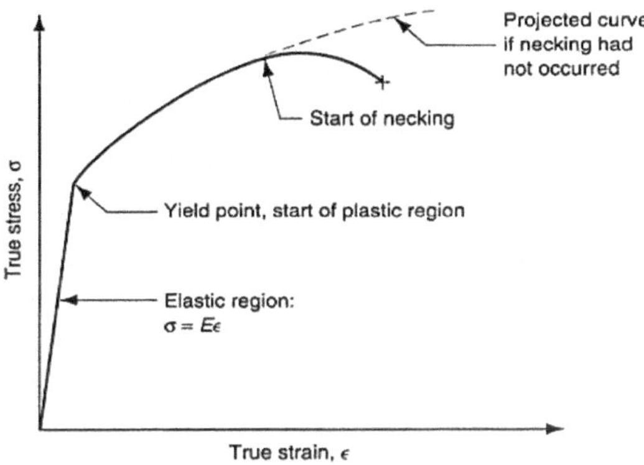

Figure 3.1

Materials Used for Cold Working Process

1. Low and medium carbon steel.
2. Copper and light alloys.
3. Materials like Al, Mg and Titanium.

3.5. Classification of Cold Working Process

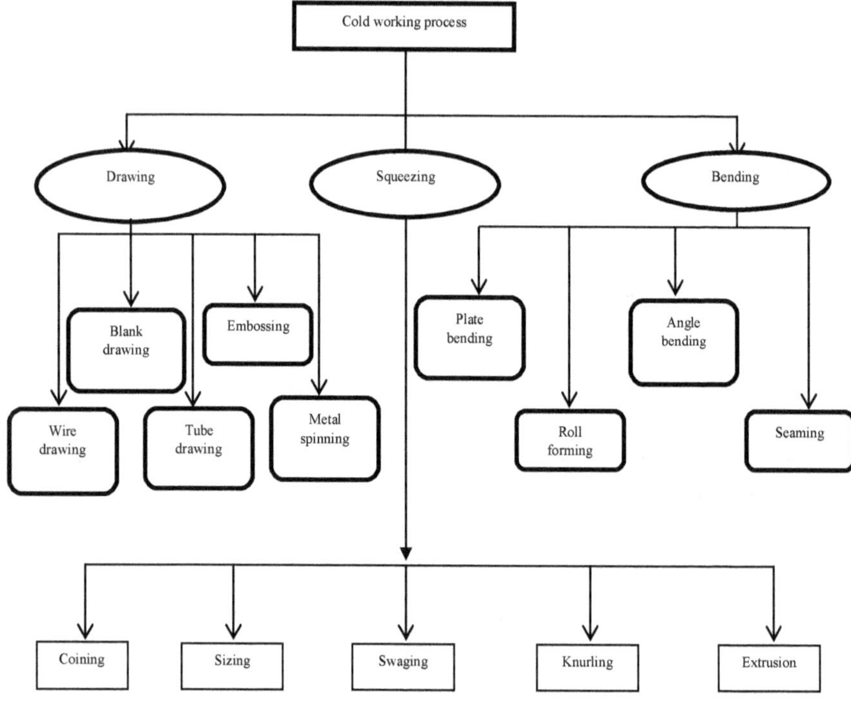

Figure 3.2

3.6. Cold Working Process

Cold Drawing

The cold drawing is used to draw the tubes, rods and wires. The classification of cold drawing is

a) Blank drawing
b) Tube drawing
c) Embossing

d) Wire drawing

e) Metal pinning

Blank Drawing

Cutting a flat shape from the metal is called blanking. The portion which is removed from the metal is called 'blank'. It is done by using the punch and die.

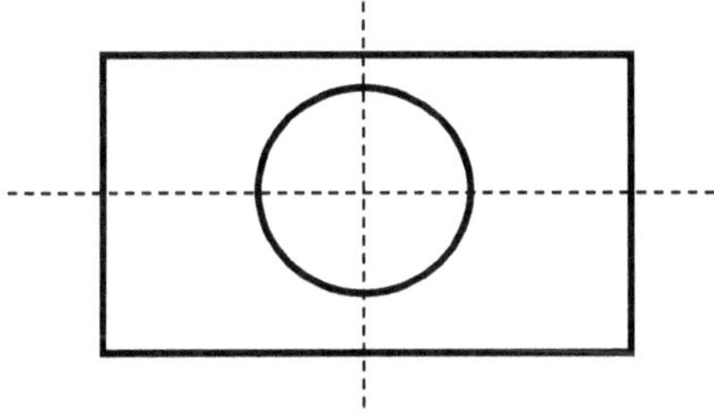

Figure 3.3

Tube Drawing

It will be discussed in detail in coming chapters.

3.7.   Embossing

- This is the process of making raised or projection design on the surface of the metal with is corresponding relief on the other side.

- It is used die set which consists of die and punch with desired shape.

- It is very useful for producing nameplates tags and designs on the metal.

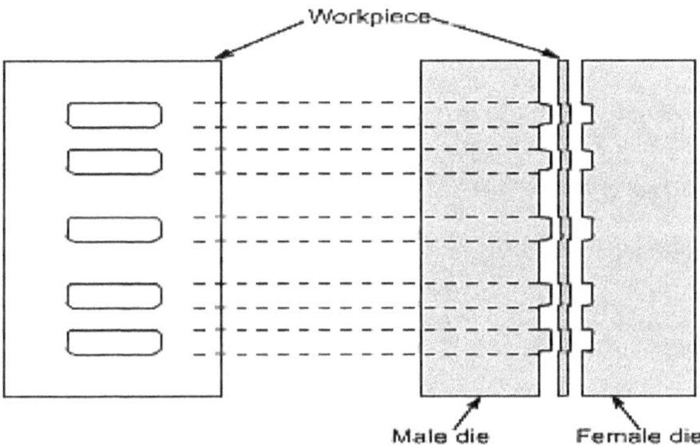

Figure 3.4: Embossing

Wire Drawing

It will be discussed in detail in article.

3.8.   Metal Spinning (Cold Drawing)

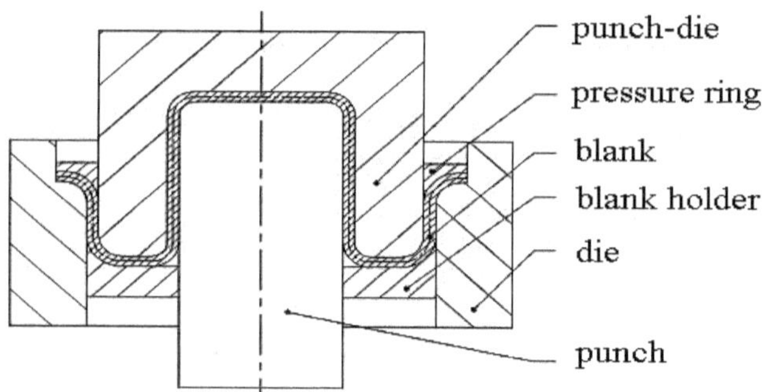

Figure 3.5: Wire Drawing

It is similar to hot spinning. The metal is pressed on the chuck attached to the lathe spindle. An adapter fitted in the barrel of the tailstock holds the work against the form block.

This process of spinning is used for

1.   Limited production
2.   Keeping the tool cost low.

3.9.   Coining

Process where metal while it is confined in a closed set of die; used to coins, medals and other products where extract size and fine details are required, and thickness varies about well-defined average.

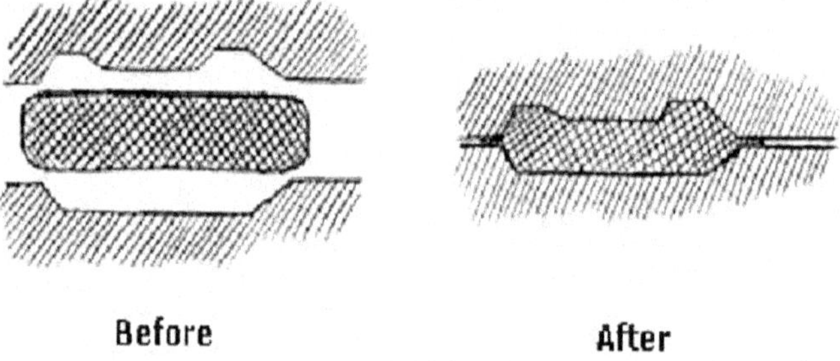

Figure 3.6: Coining

Sizing

This operation in cold working is used to size the metal into the required shape.

Swaging

Process that reduces/increases the diameter, tapers, rods or points round bars or tubes by external hammering.

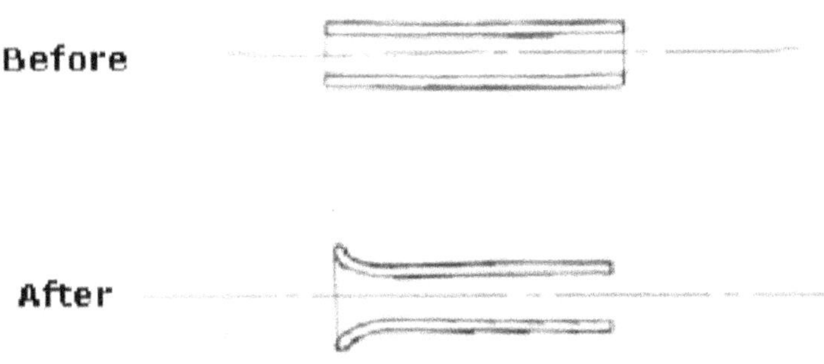

Figure 3.7: Swaging

3.10. Knurling

Knurling is performed on a lathe with hardened rolls. The surface of the rolls is a replica of the profile to be generated. The rolls are pressed radially against the rotating work-piece. It is used to make grip on the handles in a machine.

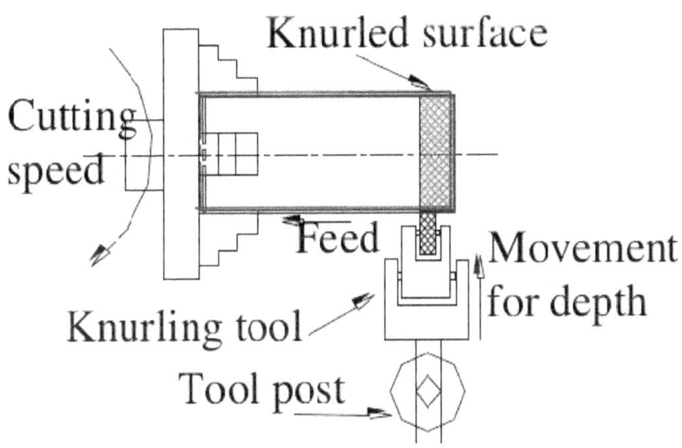

Figure 3.8: Knurling

Extrusion

- Process which is commonly used to make collapsible tubes such as toothpaste tubes, cans usually using soft material such as aluminium, lead and tin.
- Usually a small shot of solid material is placed in the die.

Bending

Cold bending is employed for bending into desired shapes various stock materials like rods, wires, bars, pipes, tubes and various structural shapes. It is usually performed in many stages. Well-designed fixtures are also used where mass bending of such components is required.

The bending operation is mainly classified into:

a) Plate ending
b) Rll forming
c) Angle bending, and
d) Seaming

Plate Bending

In this type of operation, large size plates are bent by using the bending machine.

Roll Forming

The bending machine carries three rolls. Mnog, these, two are fixed and the third one is adjustable. But the diameters of all the rolls are same.

By adjusting the position of the adjustable roll, the plates or sheets can be bent in different curvatures.

Angle Bending

Shapes like angles, ovals and circles are made. The angle bending is similar to hot angle-bending operation.

Advantages of Cold Working

- It is widely applied as a forming process for steel.
- Better surface finish is being obtained.
- Thin material can be obtained.
- It is more suitable for mass production.
- This process provides higher dimensional accuracy.

Limitations

- The surface finish may be poor.
- Close tolerances cannot be achieved.
- Stress formation in the metal during cold working is higher.

3.11. Comparison Between Hot Working and Cold Working

S.NO	Hot working	Cold working
1	Working above recrystallization temperature.	Below recrystallization temperature.
2	Harden the metal	No hardening
3	Surface finish is not good	Good surfae finish can be obtained
4	New crystal are formed after hot working	No crystallization
5	Internal stress is not formed	Stesss formation in the metal well occur
6	Elongation of metal takes place	Elongation decreases
7	Imprurities are removed from the metal	Imprurities are not removed
8	Large size metals also deformed	Limited to size
9	Blowholes, cracks get welding during hot working	Ductility is obtained during cold working and it is useful for machining process

3.12. Forging Process

Synopsis

- Introduction of forging process
- Open – Die Hummer Forging
- Impression – Die Drop Forging (Closed – Die Forging)
- Press Forging
- Upset Forging
- Roll Forging

Introduction of Forging Process

Forging is a process in which material is shaped by the application of localized compressive forces exerted manually or with power hammers, presses or special forging machines. The process may be carried out on materials in either hot or cold state. When forging is done cold, processes are given special names. Therefore, the term forging usually implies hot forging carried out at temperatures which are above the recrystallization temperature of the material.

Forging is an effective method of producing many useful shapes. The process is generally used to produce discrete parts. Typical forged parts include rivets, bolts, crane hooks, connecting rods, gears, turbine shafts, hand tools, railroads, and a variety of structural components used to manufacture machinery. The forged parts have good strength and toughness, they can be used reliably for highly stressed and critical applications.

Open–Die Hummer Forging

It is the simplest forging process which is quite flexible but not suitable for large scale production. It is a slow process. The resulting size and shape of the forging are dependent on the skill of the operator.

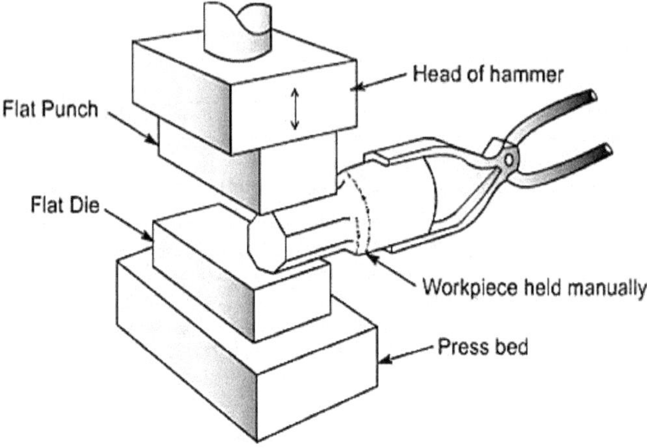

Figure 3.9: Open Die Hammer Forging

Impression – Die Drop Forging (Closed – Die Forging)

The process uses shaped dies to control the flow of metal. The heated metal is positioned in the lower cavity and on it one or more blows are struck by the upper die. This hammering makes the metal to flow and fill the die cavity completely. Excess metal is squeezed out around the periphery of the cavity to form flash. On completion of forging, the flash is trimmed off with the help of a trimming die.

Most impression–die sets contain several cavities. The work material is given final desired shape in stages as it is deformed in successive cavities in the die set. The shape of the cavities cause the metal to flow in desired direction, thereby imparting desired fibre structure to the component.

Drop Forging

In drop forging, impression dies called closed dies used. The upper die is fitted on the ram and the lower die is fitted on the anvil. Both the dies have impressions. Two rollers are fitted on the board when both rells rotated opposite to each other. When the rolls are released, the ram will falls down and, producing a working stroke. A single blow of press makes small and simple parts and large complicated shapes are made by number of steps

Applications

It is used for making

1. Spanneer
2. Automobile
3. Machine parts

3.13. Press Forging

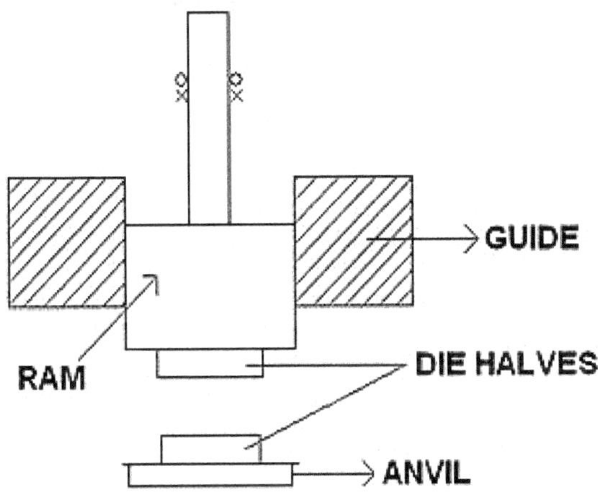

Figure 3.10: Press Forging

Press forging, which is mostly used for forging of large sections of metal, uses hydraulic press to obtain slow and squeezing action instead of a series of blows as in drop forging. The continuous action of the hydraulicpress helps to obtain uniform deformation throughout the entire depth of the workpiece. Therefore, the impressions obtained in press forging are more clean. Press forgings generally need smaller draft than drop forgings and have greater dimensional accuracy. Dies are generally heated during press forging to reduce heat loss, promote more uniform metal flow and production of finer details.

Application

1. Connecting rod

2. Spanner

3. Machine components

Hydraulic presses are available in the capacity range of 5 MN to 500 MN but 10 MN to 100 MN capacity presses are more common.

3.14. Upset Forging

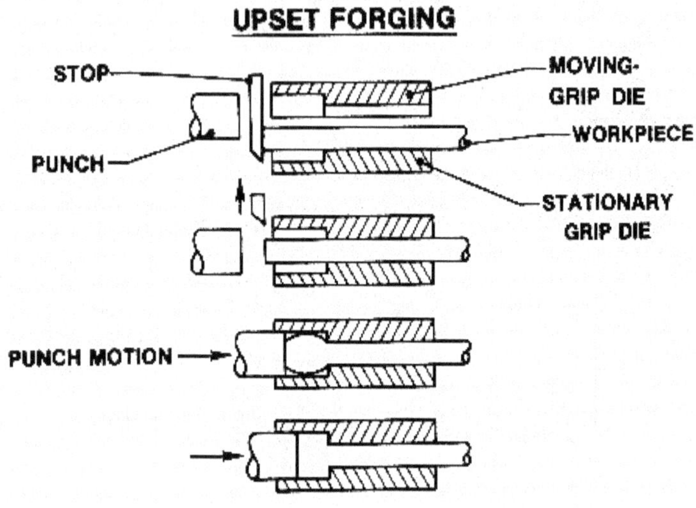

Figure 3.11: Upset Forging

Upset forging involves increasing the cross–section of a material at the expense of its corresponding length. Upset–forging was initially developed for making bolt heads in a continuous manner, but presently it is the most widely used of all forging processes. Parts can be upset – forged from bars or rods upto 200 mm in diameter in both hot and cold condition. Examples of upset forged parts are fasteners, valves, nails, and couplings.

The process uses split dies with one or several cavities in the die. Upon separation of split die, the heated bar is moved from one cavity to the next. The split dies are then forced together to grip the and a heading tool (or ram) advances axially against the bar, upsetting it to completely fill the die cavity. Upon completion of upsetting process the heading tool comes back and the movable split die releases the stock.

Upsetting machines, called upsetters, are generally horizontal acting. When designing parts for upset–forging, the following three rules must be followed.

1. The length of unsupported bar that can be upset in one blow of heading tool should not exceed 3 times the diameter of bar. Otherwise bucking will occur.

2. For upsetting length of stock greater than 3 times the diameter the cavity diameter must not exceed 1.5 times the dia of bar.

3. For upsetting length of stock greater than 3 times the diameter and when the diameter of the upset is less than 1.5 times the diameter of the bar, the length of un – supported stock beyond the face of die must not exceed diameter of the stock.

3.15. Roll Forging

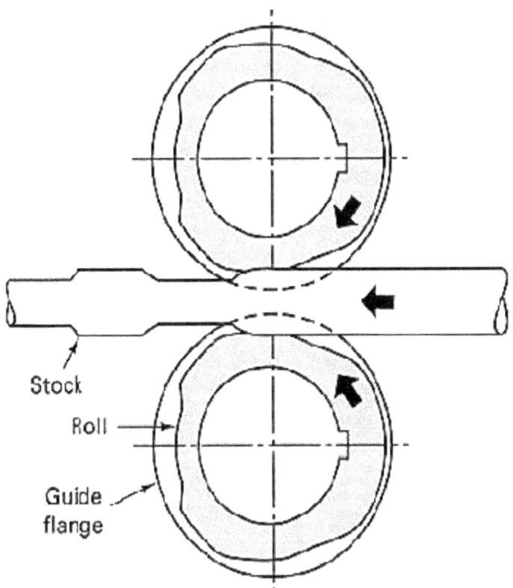

Figure 3.12: Roll Forging

This process is used to reduce the thickness of round or flat bar with the corresponding increase in length. Examples of products produced by this process include leaf springs, axles, and levers. The process is carried out on a rolling mill that has two semi–cylindrical rolls that are slightly eccentric to the axis of rotation. Each roll has a series of shaped grooves on it. When the rolls are in open position, the heated bar stock is placed between the rolls. With the rotation of rolls through half a revolution, the bar is progressively squeezed and shaped. The bar is then inserted between the next set of smaller grooves and the process is repeated till the desired shape and size are achieved.

Comparison between Press Forging and Drop Forging

S.NO	Press forging	Drop forging
1	Alignment of dies os easier	Alignment of dies is very difficult
2	Quite operation	Noisy operation
3	Structural quality of the product is superior	Not superior
4	Simple maintenance	Maintenance is difficult
5	Strokes and ram speed is high	Stroke and ram speed is low
6	One stroke is enough for making one component	More than one storke is required for making one component
7	Faster process	Slow process
8	High out put	It required skilled operators

3.16. Types of Forging Machine

Power Hammers

The various types of hammers are used to perform forging operations. They are

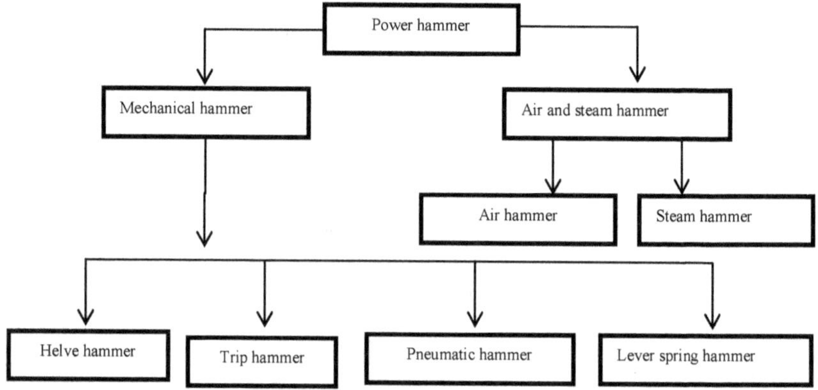

Figure 3.13

3.17. Air and Steam Hammers

Air Hammers

This type of hammer is operated by using steam or air. Steam-air hammers are classified into:

1. Single acting hammers
2. Double acting hammers

In single acting hammers, the air pressure is only used to lift the ram whereas in double acting hammer the air or steam pressure is used to lift the ram as well as impact the work piece.

Steam Hammer

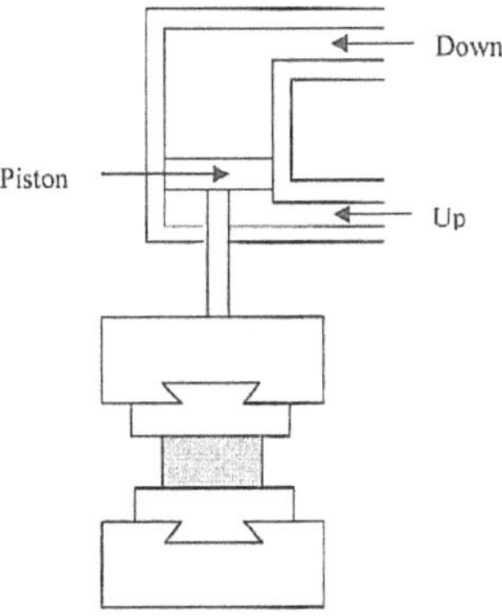

Figure 3.14: Steam Hammer

Single acting steam hammer are mainly used for light work and double acting hammers are used for heavy work. These hammers basically differ from the pneumatic hammer in that the compression of steam or air takes place separately and not within the hammer. The steam or air is supplied to the hammer under pressure.

3.18. Mechanical Hammers

Helve Hammer

It is used in general engineering work and the size can be easily changed. The main part of the helve hammer is wooden helve and the pivoted end. The adjustable eccentric between the pivot and the hammer end operatates the helve.

Trip Hammers

The vertical reciprocating ram is the main part of the trip hammers and it is actuated by toggle. The stroke length varies from 175 to 400 blows/min for small and large hammers.

Level Spring Hammer

It has constant lift and variable striking power. The main advantage of this hammer is it will increase the operating speed with increasing number of strokes. An elastic and is used to operate the ram. It is suitable for small forgings. The stroke length varies from 40 to 200 blows/min.

3.19. Pneumatic Hammers

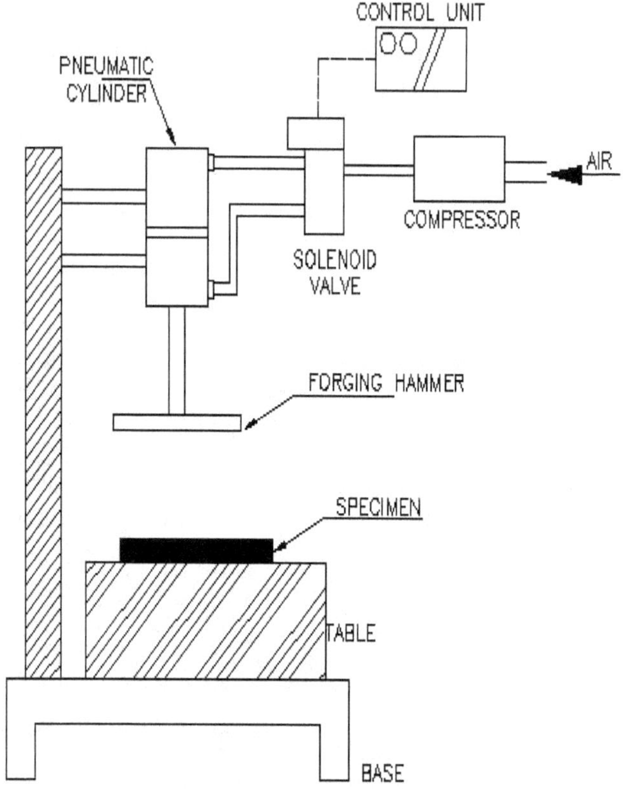

Figure 3.15: Pneumatic Hammer

The pneumatic hammers are having two cylinders.

1.   Compressor cylinder

2.   Ram cylinder

In this type of hammers, the compressor cylinder compresses the air and delivers it to the ram cylinder. This compressed air pressure is used to actuate the ram cylinder piston. The blows ranges in the pneumatic hammers varies from 70 to 190 blows/min.

The air distribution is made by rotayr valves with ports. So, the ari passes through the cylinder above and below the piston while operating the pneumatic hammers.

3.20. Power Presses

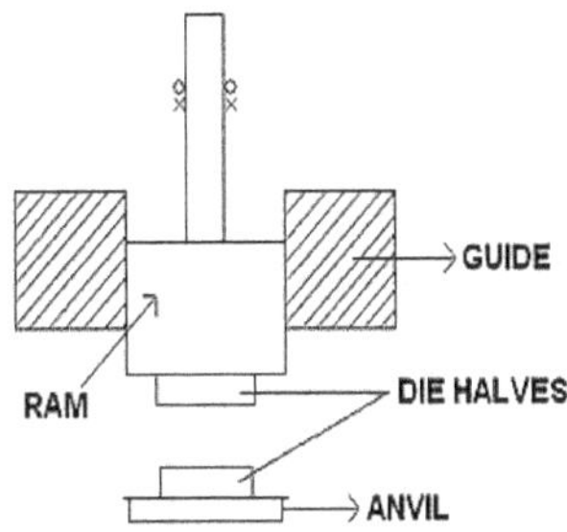

Figure 3.16: Power Press

Forging of parts by passes involve slow squeezing of plastic metal in closed impression dies instead of applying repeated severe blows by hammers. The production of forgings is by hammers because the whole operation is completed in a single squeezing action. The power presses are usually hydraulic type. The main parts of the power presses are accumulator and distributor. The pressurized waater from the pump is delivered on the press through accumulator and distributor. The ram is forced doown upon the meterial to be forged in the forward stoke. For the return stroke of the press relatively lower water pressuure, but a large volume of water per unit time is required to accelerate the ram. The presses ranging from 300×10^3 to 1000×10^3 kg are used for very large capacities. Hydraulic presses are used for heavy work and mechanical presses for light work.

3.21. Typical Forging Operations

Synopsis

- Upsetting
- Drawing down
- Punching
- Bending
- Cutting
- Forge welding
- Piercing
- Swaging

- Flattering
- Fullering

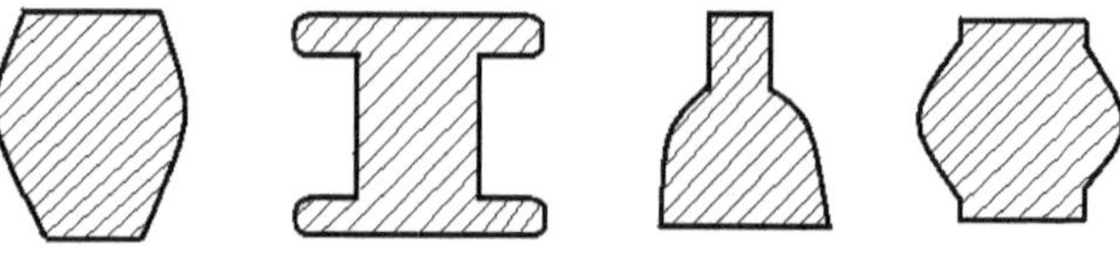

Figure 3.17: Up Setting

This process is called as hot heading. In this operation, the metal is heated at one end and it is rest on the anvil and force is applied on the other end by using hammer. So, this force will increase the cross sectional area and decrease the lengh. This operation of reducing catoss sectional area is known as upsetting.

The equipment used in this operation is called upsetter. The dies used on these machines are so designed that the complete operation is performed in several stages and the final shape attained gradually. The operation is performed with the help of a die and a punch. The various types of upsetting processes are shown above.

3.22. Drawing Down

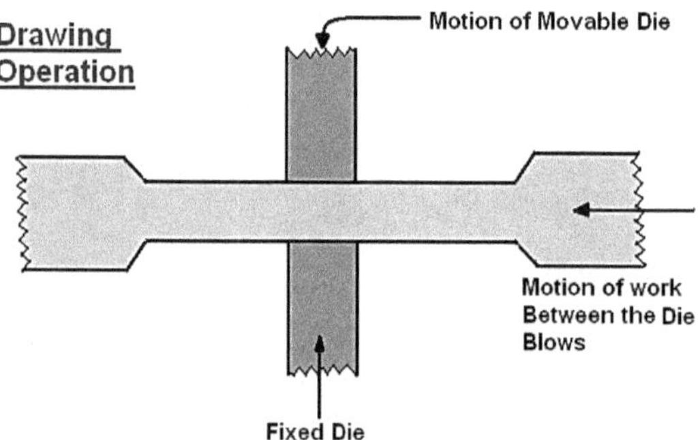

Figure 3.18: Drawing Operation

This process is also known as drawing out. In this process, the lengh of the metal increases and the cross-sectional area decreases. First, the metal is heated to a required lengh. It means, the portion which is not required to draw may be cooled. Then the bar is placed on the anvil to draw rhe bar by using fuller and hammer. It is exactly a reverse process to that of upsetting operation. The desired effect the peen of a hammer, a set of fullers or a pair of swages.

3.23. Punching

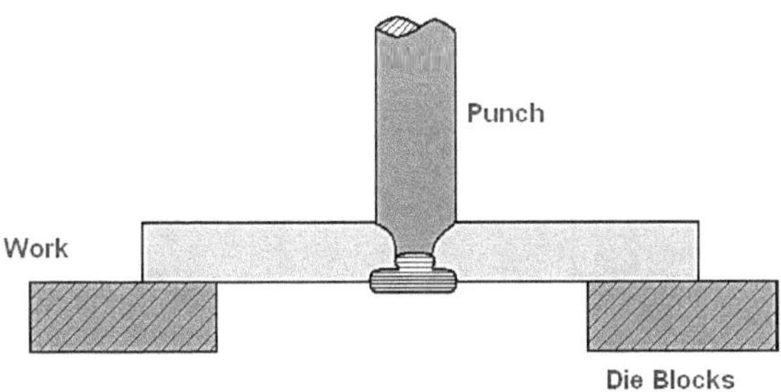

Figure 3.19: Punching

Punching is defined as making of a hole in a given job. The principle operation of this process is placing the heated job over a correct hole of the swage of die and forcing the punch into it by hammer. The work piece is initially heated tp nearly white heat and then placed flat on the anvil face. If a small hole is to be produced, the second stage of the operation can be perfomed by placing the work on the anvil face.

Bending

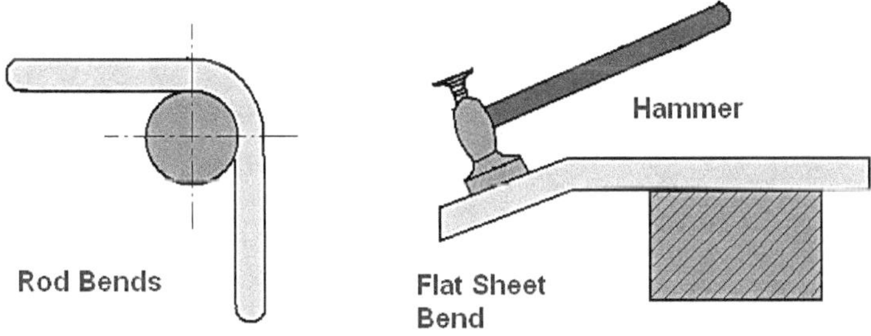

Figure 3.20: Bending

Bending is one very common sheet metal forming operation used not only to form shapes like seams, corrugations, and flanges but also to provide stiffness to the part (by increasing its moment of inertia). Shapes like angle, ovels, and circles etc can be done by this method. Before going to this operation the job is heated in the appropriate portion. Then the jop is placed over the anvil and the force is applied through the hammer. So, it provides an extra material at that particular place which compensates for the elongation of the outer surface due to hmmering during bending.

Cutting

Removel of excess metal from the work or for making pieces from a bar stock is known as cutting process. In this a pair of blades is used to cut the metal for the required shape.

3.24. Forge Welding

Joining the two work pieces by forging operation is called forge welding.

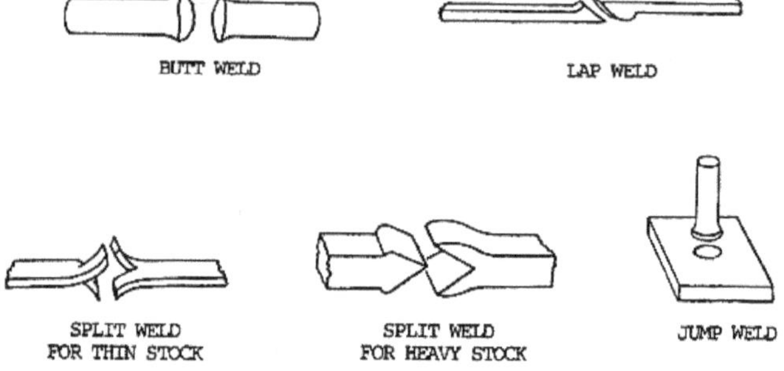

Figure 3.21: Forge Welding

In this process, the work pieces are heated and cleaned thoroughly before going to the weld and the force is applied to the pieces by hammer blow to join together. There are three district types of welded joints in common use.

1. Butt weld
2. Scarf weld
3. 'V' weld

3.25. Piercing

Making is blind are through hole with the help of a punch in the metal.

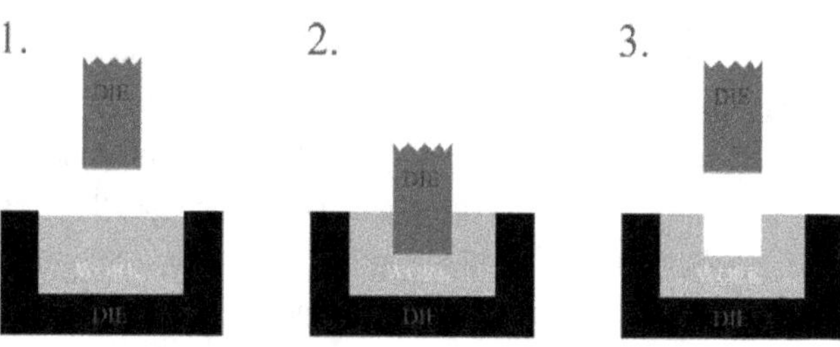

Figure 3.22: Piercing

Piercing may be followed by punching to produce a hole in the part. Piercing is also performed to produce hollow region in forgings using side auxiliary equipment. Piercing force depends on the punches cross a sectional area and tip geometry, strength and friction of the material.

3.26. Swaging

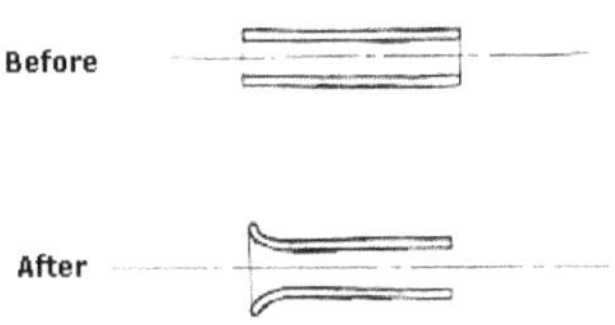

Figure 3.23: Swaging

Reducing the cross sectional of the area is known as swaging.

3.27. Flattering

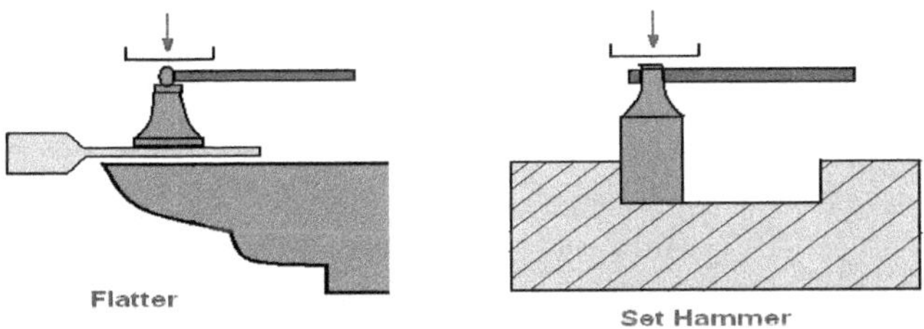

Figure 3.24: Flatting and Setting Down

It is used to flat the stock and that the stock is fitted properly in the closed die.

3.28. Fullering

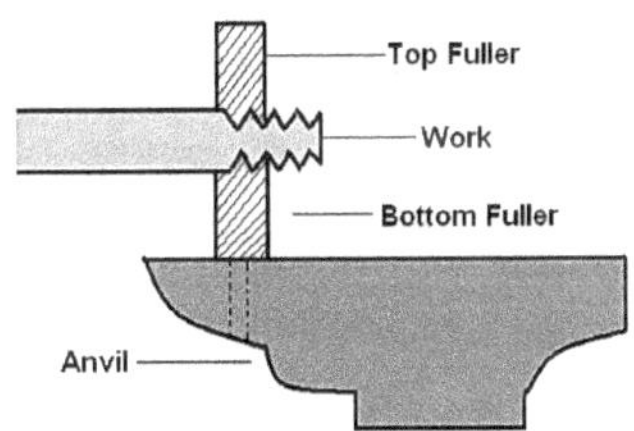

Figure 3.25: Fullering

Reducing the stock and increasing length of the workpieceBy appling pressure on it.

3.29. Rolling of Metal

Synopsis

- Introduction of Rolling of metals
- Flat strip rolling
- Classification of rolling based on numbers of rolls

Introduction of Rolling of Metals

Deforming the metal into semi finished condition by passing the metal piece between two rollers. Rolling is done by both hot and cold working.

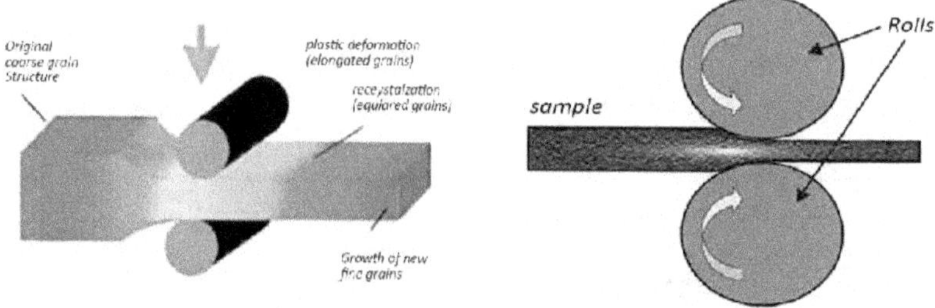

Figure 3.26: Hot Rolling

Figure 3.27: Cold Rolling

In a hot rolling, the metal is heated to a plastic state and it is passed in between the two rollers which are operated in the opposite direction whereas in cold rolling the metal is not heated and it is retained the given shape by the action of the rolls.

3.30. Flat Strip Rolling

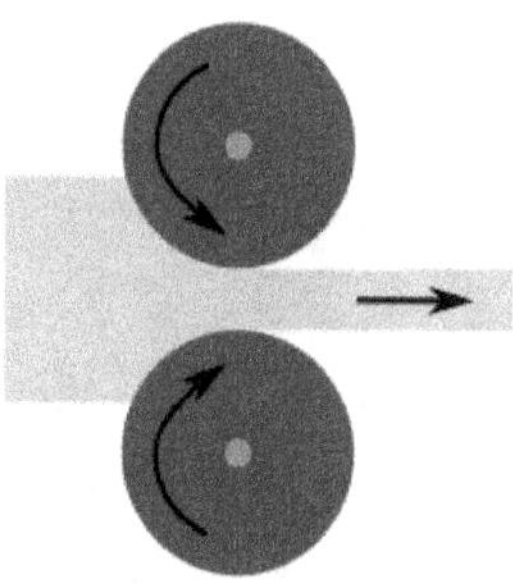

Figure 3.28: Flat Strip Rolling

In a rolling plate and sheet with high width to the thickness ratio. The width of the material remains constant during in this process for example square section, the width increases considerably in the roll gap. The increased width in the rolling is called as spreading for calculating rolling force the width is taken as average width. The diagram of flat strip rolling as shown in fig. the thickness of strip h_0 is reduce to h_f by a pair of rotating wheels. The velocity of strip increases to v_0 to v_f. since the surface of the roll is constant in roll gap L. The frictional forces which are acting on the flat strip is also given in fig. and the roll force and power requirement for the rolling is given by fig.

1) Roll force

F = Lw Y$_{avg}$

Where, L - Roll strip contact length

W -Width of the strip

Y$_{avg}$ - Average true stress.

2) Power per roll

Power = $2\pi FLN/6000$ kw

Where, F = Force in Newton

N = speed of the roll in rpm.

3) Roll strip contact length (L)

$L = \sqrt{R(h_0 - h_f)}$

Where, $h_0 - h_f$ = Difference between initial and final thickness

R = Roll radius

Classification of rolling based numbers of rolls:

3.31. Two High Rolling Mill

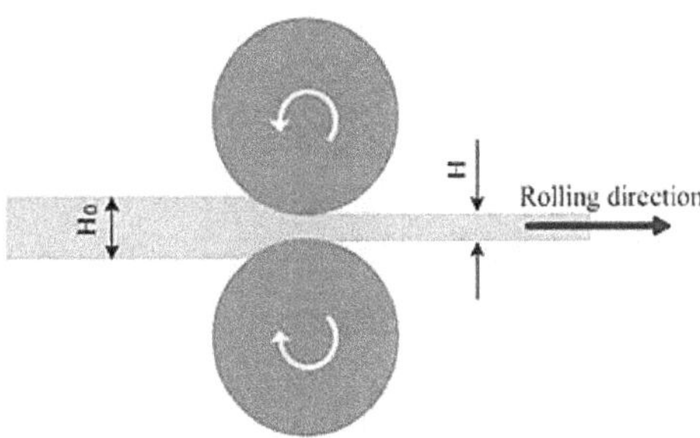

Figure 3.29: Rolling Process

In two rollers, both the rolls are rotate in a constant direction about the horizontal axis. It reduction of the cross sectional area of the stock and increasing the length. The method is most suitable for both hot and cold working.

3.32. Three High Rolling Mill

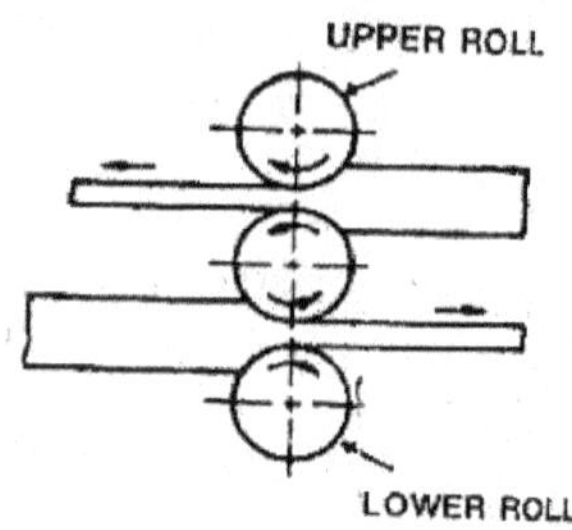

Figure 3.30: Three High Rolling Mill

As the name implies, it has three rolls which are rotate in a constant direction. The upper and lower mills are drive rolls and middle roll is rotates by friction. It is very useful form, asking billet rolling and finish rolling.

3.33. Four High Rolling Mill

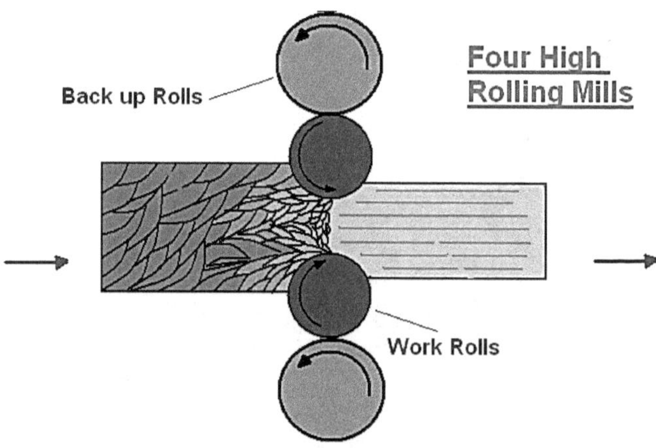

Figure 3.31: Four High Rolling Mills

It is used in reversing the mills for the rolling hot and cold rolling sheets to avoid the bending of work rolls due to low strengh and large diameter backup rolls are installed.

3.34. Multiple Roll Mills

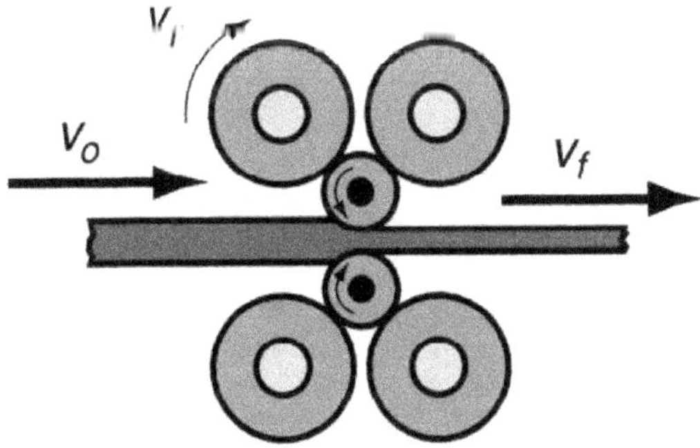

Figure 3.32: Multiple Roll Mills

In a multiple roll and the work rolls are supported by backup rolls. The work rolls are driven by driving rolls .

3.35. Universal Rolling Mill

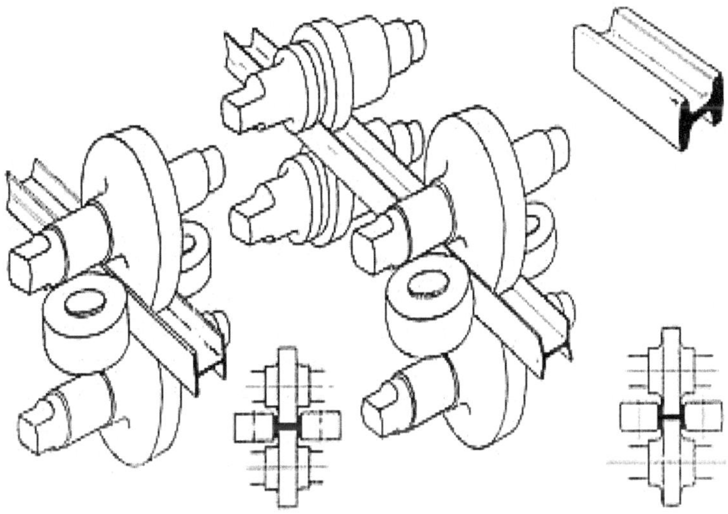

Figure 3.33: Universal Rolling Mill

The metal is reduced by both horizontal and vertical rolls. The edges of the bar is smoothen by the vertical rolls and the vertical rolls are mounted either one side.

3.36. Shape Rolling Operations

In shape rolling process, the various shapes can be produced.

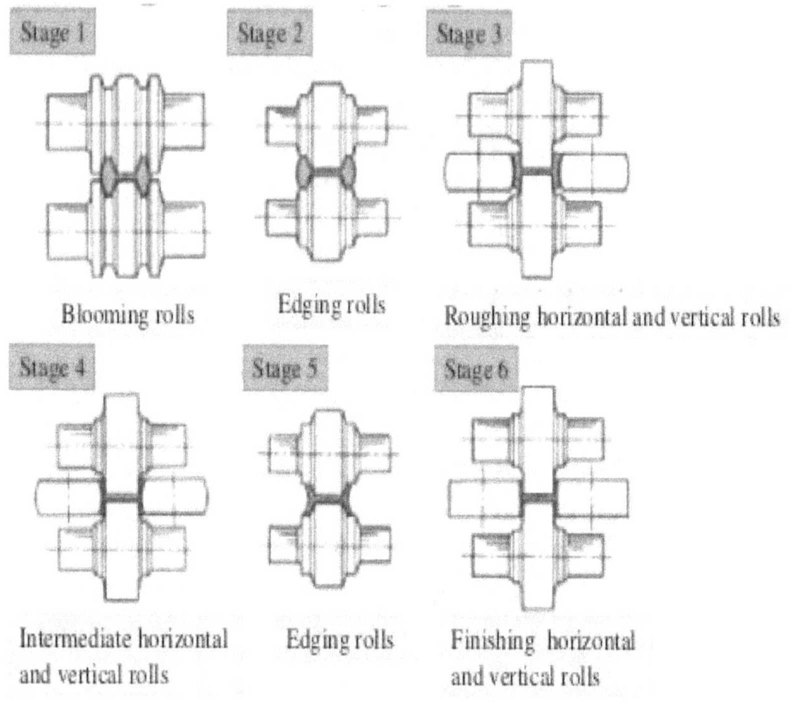

Figure 3.34: Shape-Rolling Operations

Example

Straight and long structural shapes, solid bars, I-beams, channels, railroad rails,

Types

- Ring rolling
- Thread rolling

3.37. Ring Rolling

In a ring rolling process, a thick ring is expanded in to a large diameter ring with a reduced cross section. A ring rolling process has the advantages of short production time, colse tolerance, material saving. It can be carried out it room temperature depend on size, strengh, ductility of the material.

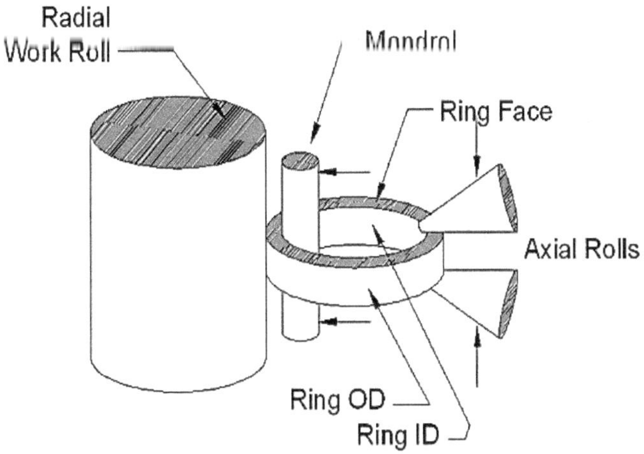

Figure 3.35: Ring Rolling

3.38. Thread Rolling

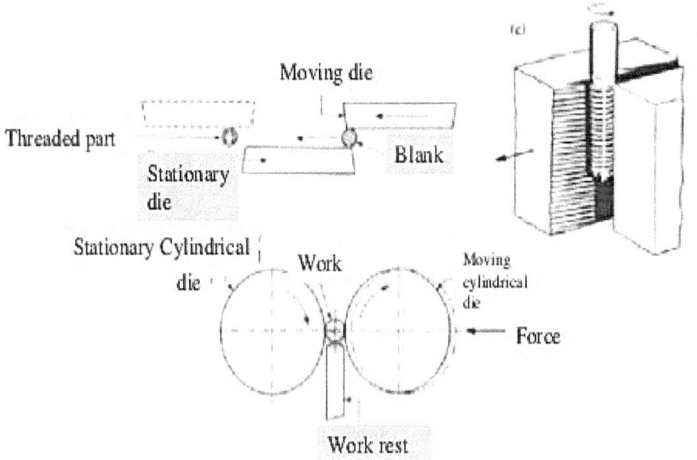

Figure 3.36: Thread Rolling

It is a cold forming process by which straight formed. The thread are formed on the rod which each stroke of pair of flat reciprocating die. It has advantages of generating thread without any loss of material. The surface finish is very good.

3.39.   Defects In Rolled Parts

There are two types

1.   Surface defects

2.   Internal structural defects

Surface Defects

1.   It include scale and crack and pits.

2.   It is due to the impurites inclusions in the original cast material

Internal Structural Defects

1.   Wavy edges

2.   Edge cracks

3.   Folds

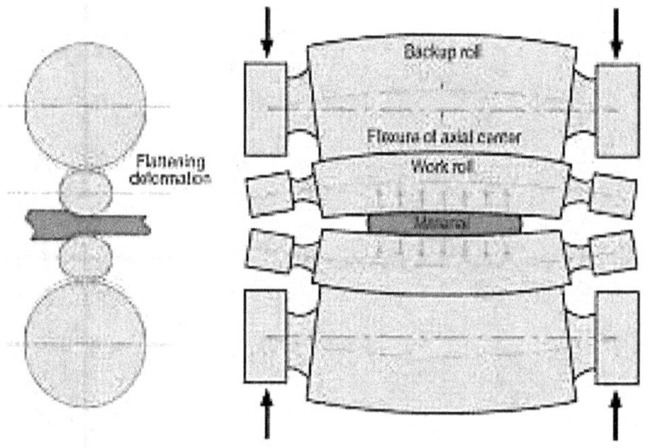

Figure 3.37: Internal Structural

The defects wavy edges due to bending of rolls. Normally the rolls act as a straight beems. If the material flow is continuous and to maintain its continuity, strains with in the material should adjust there are compressive strain on the edges and tensile strain on the center.

Other Defects in Rolling

1.   In homogeneous deformation of elements across the width.

2.   In a homogeneous deformation in the thickness section

3.   Folds

4.   Lamination

Principles of Extrusion

The heat and metal is commpressed and forced through a shape and die. The force required for the cold extrusion process is high. Metals like lead, tin, coppers are extrud in a cold condisions because of its low density.

The major extrusion defects are

1.   Surface cracks
2.   Internal cracks
3.   Pipe

Type of Extrusion

There are two types

1.   Hot extrusion
2.   Cold extrusion

3.40. Hot Extrusion

Metals like copper, lead magnesum is having low yeild strength and extrusion temperature. Hot extrusion operation is hydralically operated. The sizes are rated from 250 to 5500 tonnes of presses are used in hot extrusion process. There are two types of hot extrusion process

1.   Forward extrusion
2.   Backward extrusion

3.41. Forward Extrusion

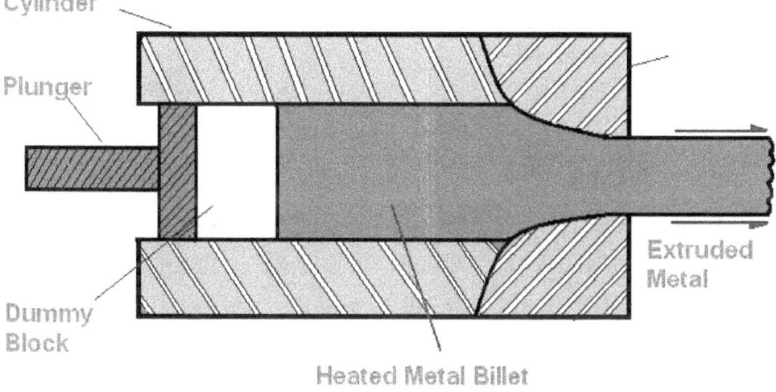

Figure 3.38: Direct or Forward Extrusion Process

The forward extrusion fig the heated billet metal is placed in press, Which is operated by a ram and a cylinder. The heatd metal billet is pushed by the ram and the applications of ram pressure. The metal first plastically fill the die. Then it is forced out through die opening and finally dcut the die face.

3.42. Backward Extrusion

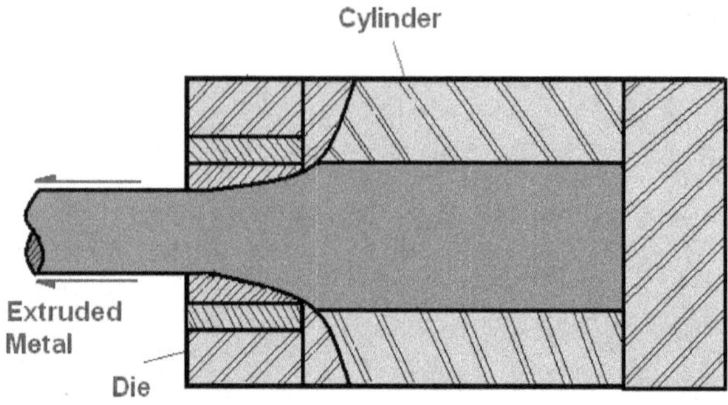

Figure 3.39: Backward or Indirect Extrusion Process

The extruded part is forced through the hollowram. The ram is operated by a hydraulic drive the heated metal is based in the die and forced applied through the hollwram. Finally cut the die face.

3.43. Cold Extrusion(Impact Extrusion)

The work material is placed between the die and the ram.

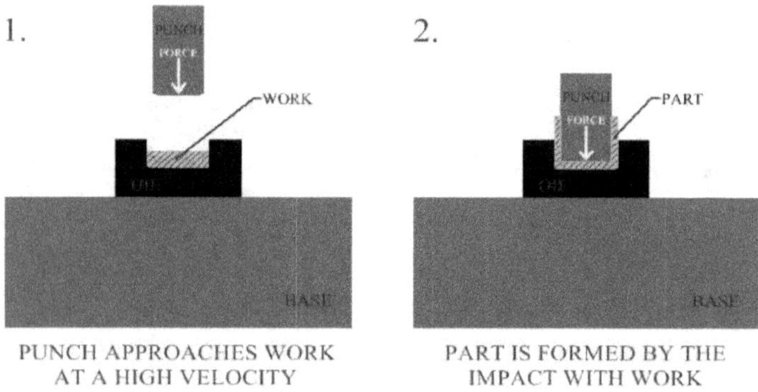

Figure 3.40: Impact Extrusion

Sudden impact is given to the ram. The metal flows plastically in the upward direction, the metals like aluminium, tin. The various items are daily use such as tube for saving cream, tooth paste, paint condenser cans and thin walled products are impact extruded. The metal flow of along the surface of the punch forming a cup shaped components.

3.44. Principles of Drawings

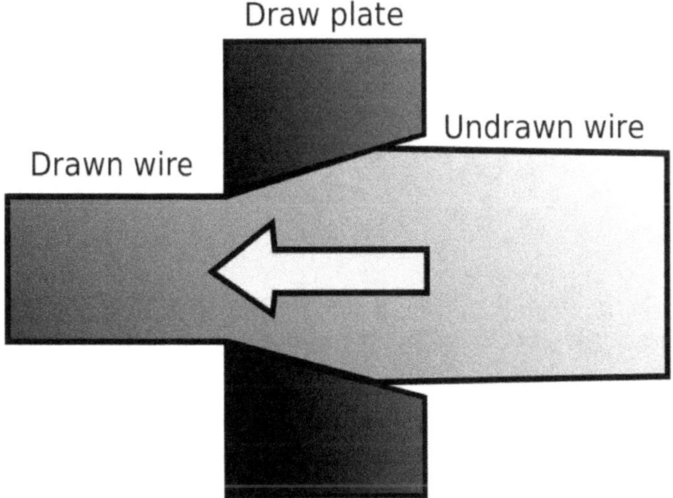

Figure 3.41: Principles of Drawings

Which is used to reduce the diameter of the metal. In a cold working process in which the material to be drawn is pulled the through a tapered hole in a die.

Application

- It is very useful to produce round rectangle square shapes
- Electrical wire and wireframes structures are made
- It is useful in vehicle and industrial machineries

3.45. Classification of Drawing

- Wire drawing
- Rod drawing
- Tube drawing
- Deep drawing

3.46. Wire Drawing

Diameter less then 16 mm has drawn in the form of wire coil.

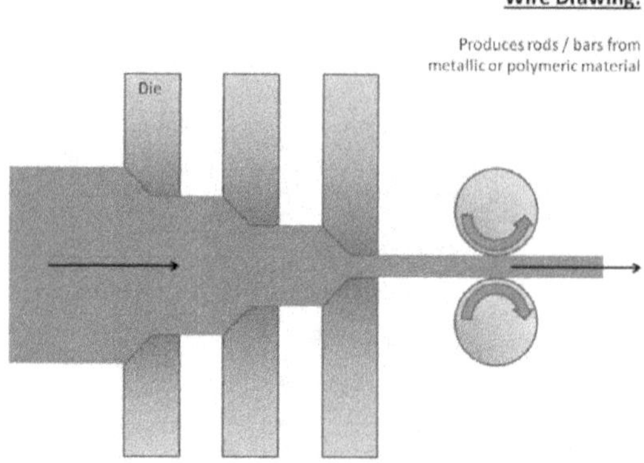

Figure 3.42: Wire Drawing

Initially the point of wire is sized so it is freely entered in the die. The sized point coming out of the die orifice is fixed on the carrige , which pulls the rod through the all zones of the die orifice. That will reduces the diameter of the rod for making the fine wire the rod is passed through the numbers of dies. Finally the wire is connected to power reel to get the wire coil.

3.47. Rod Drawing

The rod which is to be drawn should be straight and the maximum length of the rod drawn depending upon the carriage moment. The process is consists of placing the hot drawn bar through a die of which the bore size conform to the finished size of product.

3.48. Deep Drawing

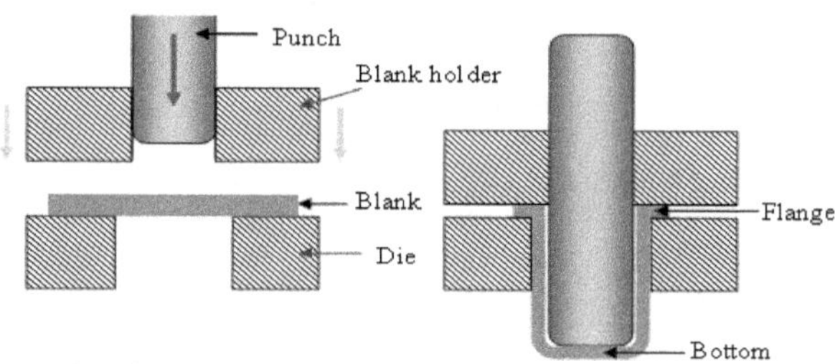

Figure 3.43: Deep Drawing

Which is used to draw the cup shaped parts from the metals, is called deep drawing.

Metal block is initially heated to a plastic state and placed on the die forces are given by using so the required shape of the cup according to the shape of the cavity produced.

3.49. Equipments Used in Drawing

The equipment for drawing is basically of two types.

1. Draw bench
2. Bull block

Draw Bench

1. It contains a single die and it is similar to a long horizontal tension-testing machine.
2. In this, the pulling force is applied by a chain drive. It is used for single length of drawing of straight rod and tube.

Bull Block

It is used to very long rod and wires in smaller cross section usually less then 13 mm. The tension in the setup provides force required for drawing the wire.

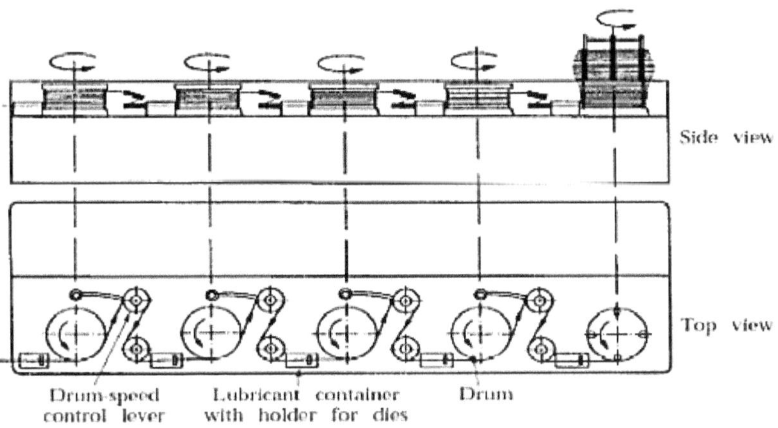

Two views of multistage wire-drawing machine that is typically used in the making of copper wire for electrical wiring.

Figure 3.44: Drawing Equipment

UNIT 3

METAL FORMING PROCESSES

PART-A (2 MARKS)

1. List out the types of forging machines
2. What are the types of rolling mills?
3. What are the four major draw backs of hot working?
4. Classify the types of extrusion
5. State any two effects produced by Cold-working
6. What are the two basic types of forging process?
7. What do you understand by forging? What are the advantages?
8. List out the forging defects
9. Classify the types of forging machines
10. State the defects in rolled parts.
11. What are the advantages of cold forming?
12. What is the purpose of piercing operation?
13. Name any four limitations of hot forging
14. Write the limitations of hot working process
15. What is the difference between stretch forming and bending?
16. What do you understand by recrystallization and recrystallization temperature?
17. What are the general advantages of forging as a manufacturing process?
18. List the functions of Back-up rollers in rolling operation?
19. Discuss in brief open die and closed die forging
20. What is the principle of impact forging?

PART-B (16 MARKS)

1. Classify the types of forging machines and explain any one
2. Explain the forward and back extrusion process
3. i. Classify the types of rolling mills and sketch them

   ii. List out various forging defects
4. i. Describe hydrostatic extrusion process.

   ii. Compare press forging and hammer forging
5. i. Explain the tube piercing process

   ii. Distinguish hot and cold extrusion process and briefly explain one in each.

6. i. Describe the principle of rolling. Write the various kinds of rolling mills along with their applications.

ii. What are the types of power hammers available and explain the pneumatic hammer with a neat sketch

7. i. Describe the difference between a bloom, a slab and a billet. Explain the features of different types of rolling process.

ii. Discuss the effects of temperature, strain rate and friction on metal forming process

8. i. Explain with a sketch, what is meant by flat strip rolling.

ii. Explain the procedure for making the head of Bolt by forging operation

9. i. Name the hand forging operation and explain briefly about them.

ii. Explain with a neat sketch of roll forging process.

10. Describe the following processes

a. Roll die forging   b. Skew rolling   c. Ring rolling

UNIT 4

SHAPING & FORMING

4.1. Sheet Metal Characteristics

Roll Forming

- Long parts with constant complex cross sections good surface
- Stretch forming
- Long parts with shallow contours suitable for low quantity

Drawing Forming

- Deep parts with relatively simple shapes
- Stamping
- It includes a variety of operation such as punching blanking coining
- labour cost is low

Rubber Forming

- Flexibility of operation low tooling cost

Spinning

- Small parts good surface

Super Plastic Forming

- Complex shapes fine detail so parts are not suitable high temperature use

Peen Forming

- equipment cost can be high

Shearing Operations

- Shearing
- Bending
- Drawing
- Squeezing
- The process of cutting straight line across a strip is known as shearing process
  1.Plastic deformation
  2.Fracture
  3.Shear

Selection Process

- Force required
- Die space
- Size and type
- Stroke length
- Method of feeding
- Shut height
- Type of operation
- Speed of operation

4.2. Fly Press

Construction of Press

- Manually operated having a cost iron frame.
- Screw provided at the top of the frame is moved in a nut
- Screw are rotated by the handle
- The punch is connected at the bottom
- The support two cast iron weights acts as a energy storage device
- An improved and heavier form of fly press is called double side press

Diagram

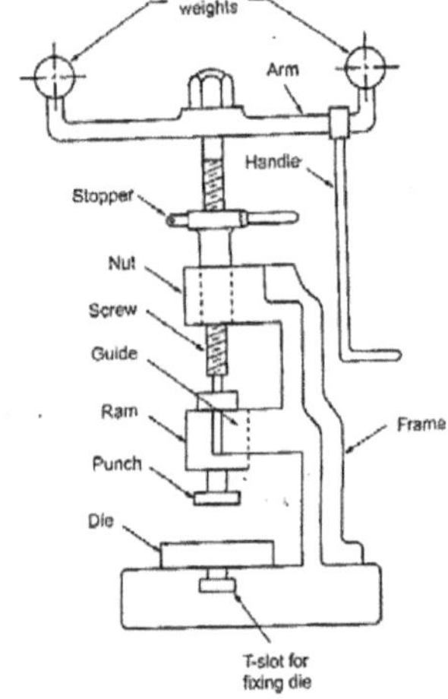

Figure 4.1

4.3.   Mechanically Operated Power Press

- An open back inclinable c shaped frame
- It allows working place between the bed and ram
- It provides easy loading

Diagram

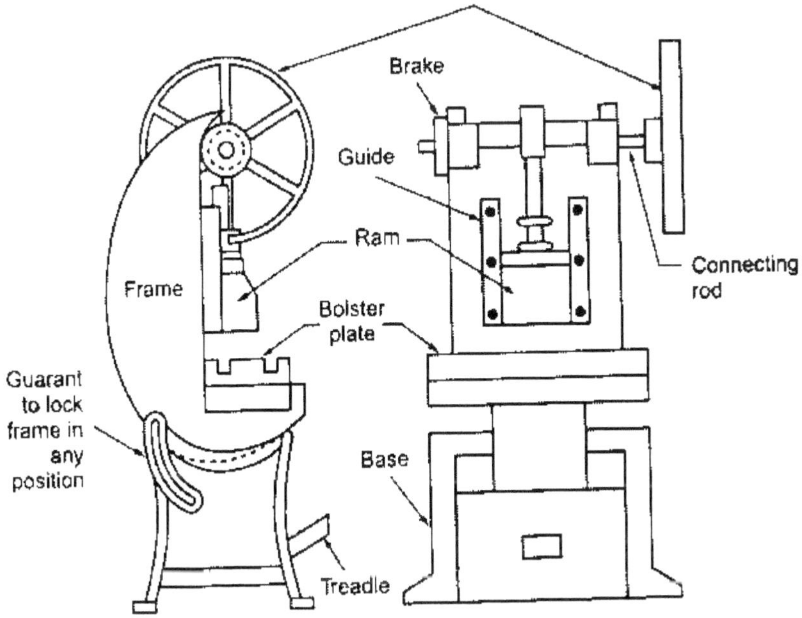

Figure 4.2: Mechanically Operated Press

- Rectangular bed
- It part of the frame having centre open

Bolster Plate

- This is nothing a thick flat thick steel plate over which press tools

Ram

- It's a reciprocating member

Knock Out

- It eject work pieces from a press tool after finishing work

Cushion

- It can be operated by oil rubber springs

Inclined Press

- The frame of the press in inclined at angle to ensure easy discharge.
- During operations the work piece can slide by gravity

Diagram

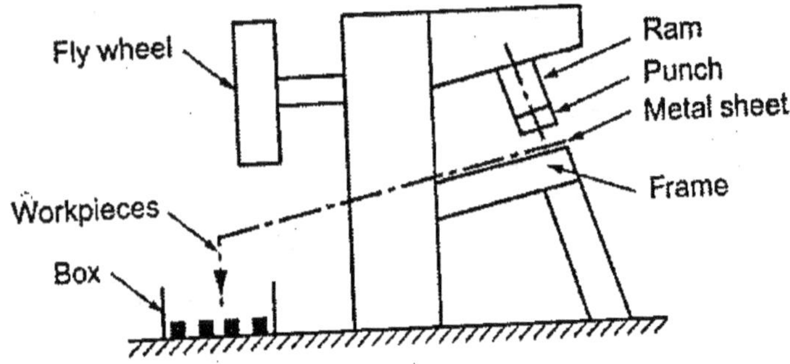

Figure 4.3

4.4. Inclinable Press

- It can be used both vertical position
- It's very much suited for higher production rate
- The work piece can slide down into the box without manual movements

Diagram

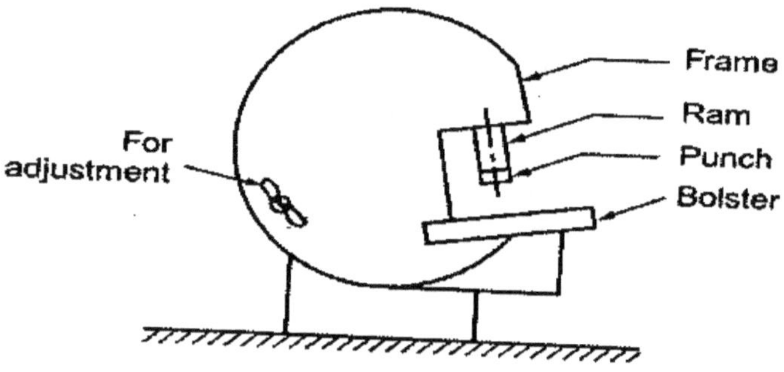

Figure 4.4

4.5. Gap Press

- Gap press has an open throat arrangement which provides clearance around the die to permit wide.

Diagram

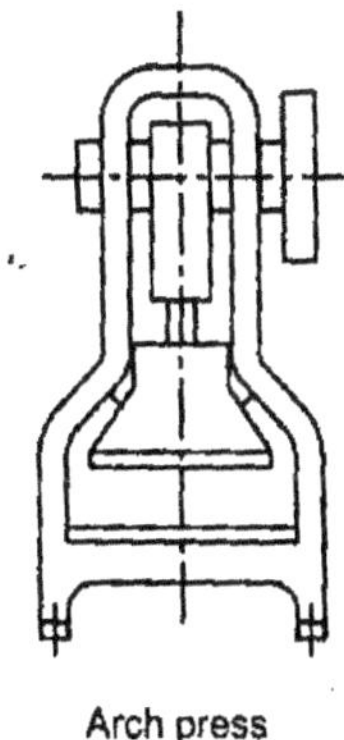

Figure 4.5

4.6. Arch Press

- The shape of frame is in the form of arch
- It is suited for blanking, trimming used in paint cans
- This type not designed for heavier work

Diagram

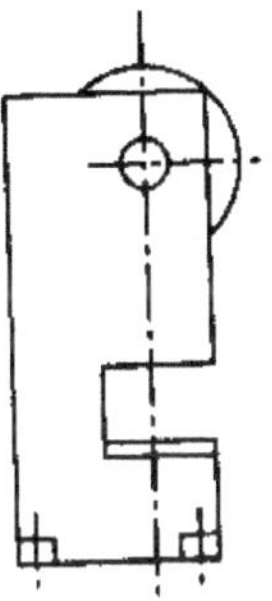

Figure 4.6

4.7.   Straight Side Press

- It is used in mechanical and hydraulic press
- High capacity presses with increased strength

Diagram

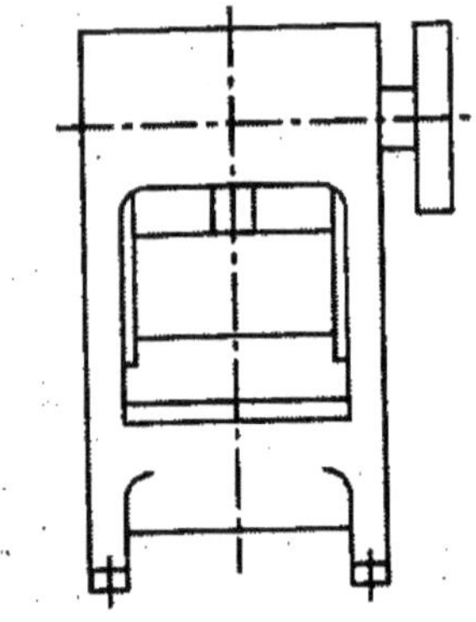

Figure 4.7

4.8.   Horn Press

- Cylindrical objects performing seaming, flanking And riveting operations

Diagram

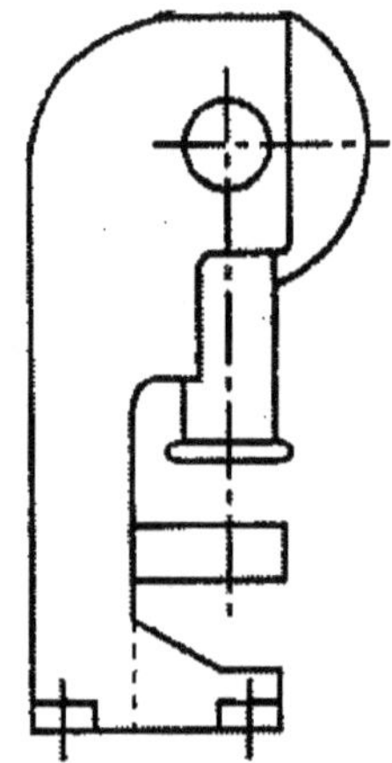

Figure 4.8

4.9. Crank Driven Press

- The ram slides moves up and down within the guide when the crank rotates

Diagram

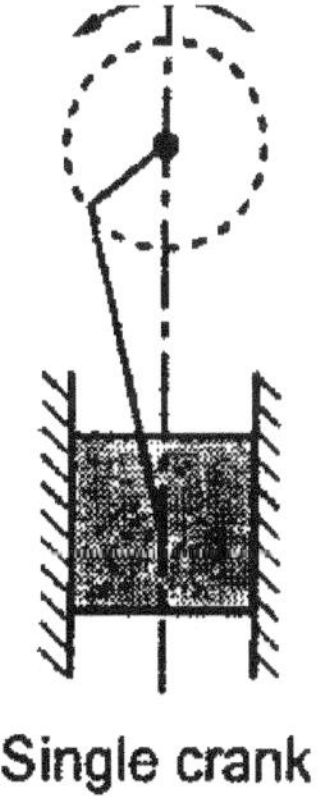

Figure 4.9

4.10. Eccentric Driven Press

- The ram slides moves up when the eccentric rotates
- It's mainly used for shorter stroke

Diagram

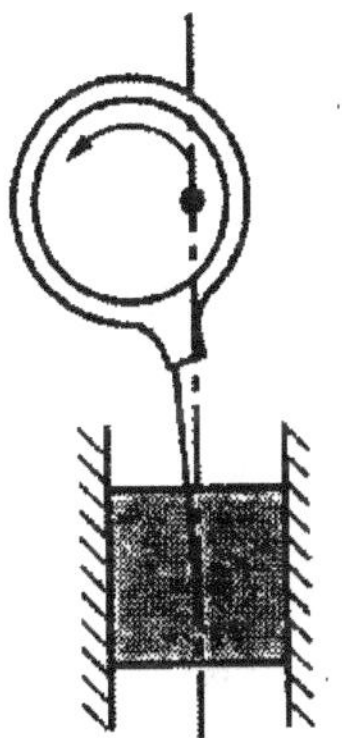

Figure 4.10

4.11. Rack and Gear Driven Press

- As the pinion gear rotates the ram connected with rack slides up and down
- To raise the ram quickly
- It very much suitable for long stroke with adjustable length.

Diagram

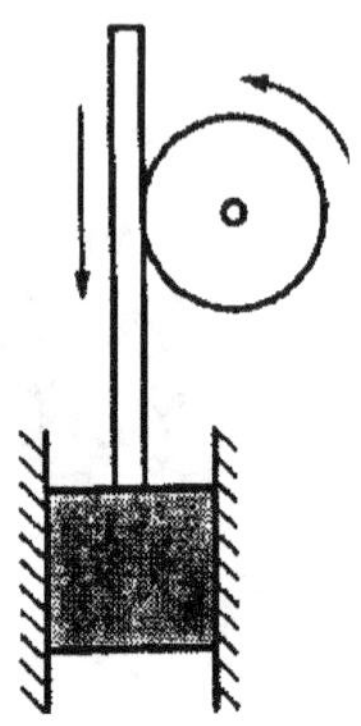

Figure 4.11

4.12. Hydraulically Driven Press

- Wherever large forces are required in forming pressing

Diagram

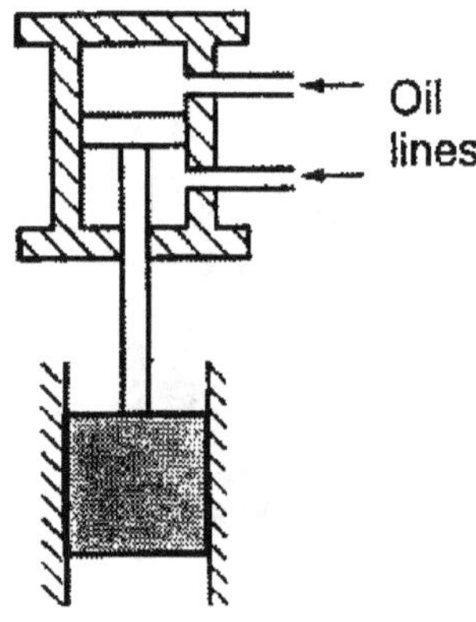

Figure 4.12

Presses with Knuckle Joint

- Where high mechanical advantage near the bottom of the stroke is necessary a knuckle joint drive is used

Presses with Toggle Joint

- It's mainly used for holding flank from its circumferences during the period.

Diagram

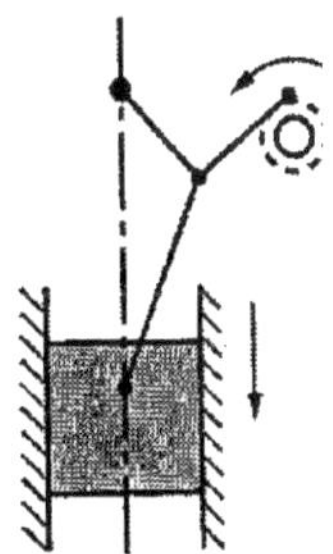

Figure 4.13

4.13. Presses with Screw Drive

- Friction disc engaged the fly wheel to slides the ram up and down
- The stored energy is used to perform the operation as required

Diagram

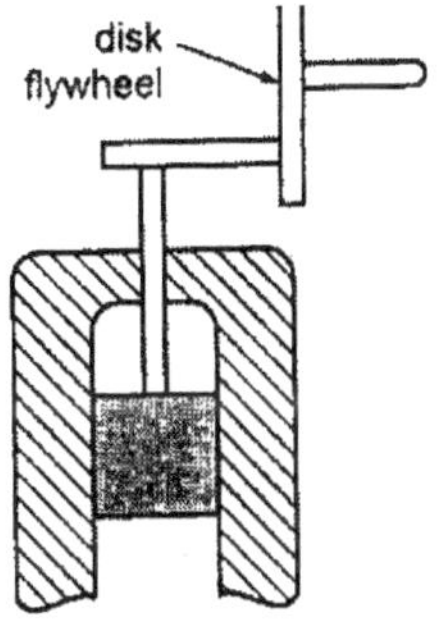

Figure 4.14

Single Action Press

- It have only one side

Double Action Press

- Two slides with two motions properly double action
- The outer slides carries the blank

Triple Action Press

- Three slides are mounted properly to ensure to three motions for triple action

4.14. Types of Cutting Dies

Progressive Dies

Synopsis

- Introduction
- Diagram
- Parts
- Working
- Merits
- Demerits

Introduction

- These die are designed perform two operations at different stages
- The stock strips is advanced through a series of action perform die operations
- The strip must move from a produce a complete work piece

Diagram

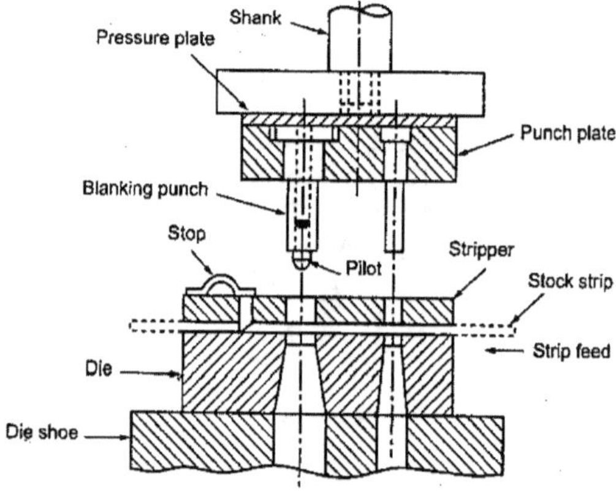

Figure 4.15

Working

The stock strip is fed into a die mechanically by a hand

- The primary stop is pushed in by hand
- When the primary stop is drawn back after the hole is pierced
- After the primary die stop is released contacts the automatic button die stop
- The pilots of the blanking punch enters the part is blanked
- While the blanking operation is performed the peering such is produces a hole
- Each stroke of the press produces a finished washer
- When establishes the progressive dies punching operation must be placed first
- Bending and forming must be done in the later stations
- The blanking and piercing punches in not be same height
- It given order to reduces the forces
- The punch will be flat and shear is given to the die

Merits

- It suitable for mass production
- Every stoke of the ram one work piece is made

Demerits

- It is complicated design of die set as compared with simple dies

Compound Dies

Synopsis

- Introduction
- Diagram
- Parts
- Working
- Merits
- Demerits

Introduction

- Two or more cutting operations
- The piercing punch is acts in the opposite direction with respect to the blanking punch.

Diagram

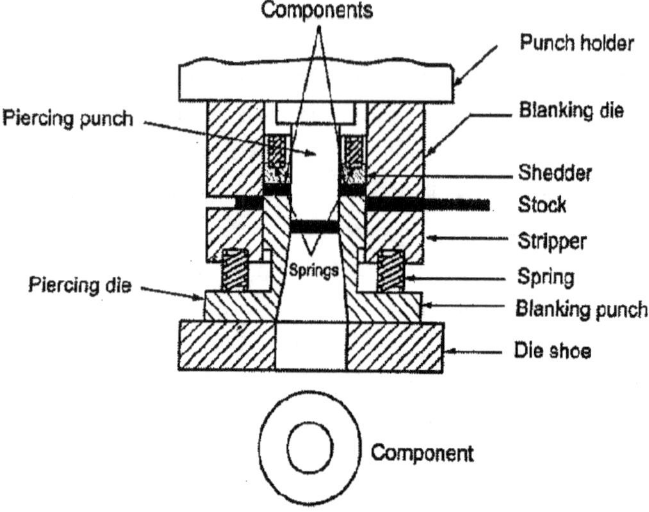

Figure 4.16

Working

- The blanking punch is also serves as the piercing die
- Its mounted on invert position
- The blanking die opening are straight because the blank not pass through the die
- During the part of the stroke hole is done on the stock the blanking operation is done
- The work piece will not be progressed in stations
- The main difference between progressive and compound dies is n numbers of washers can be produced
- It mainly used in cutting operations

Merits

- More accurate work piece can be produced
- The cost of production is very less

De merits

- Its more expensive to construct and repair
- The tonnage requirement is high

4.15. Blanking

- The cutout portion is required in this operation
- The die is made to exact size

Diagram

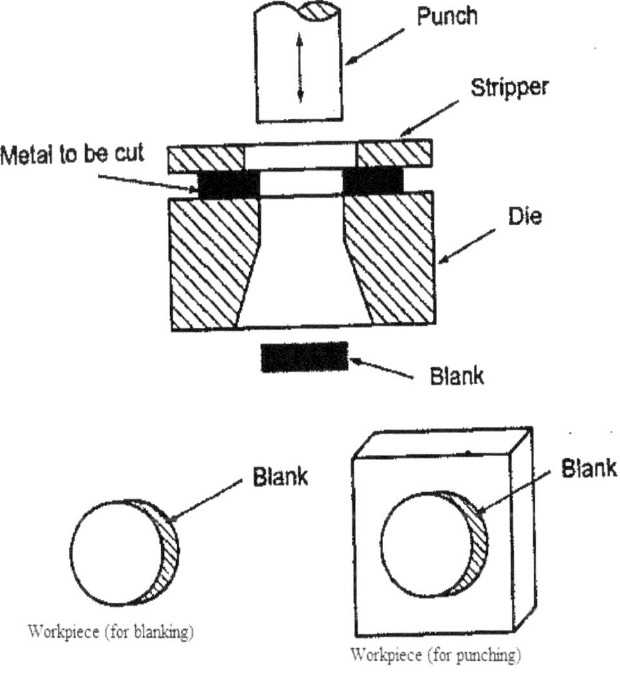

Figure 4.17

4.16. Piercing

- The cut-out portion is considered as waste and left out portion is the required part
- Basic shearing operations;
- Cutting operation
- Forming operation

Cutting Operations

- The work piece is stressed beyond its ultimate strength
- The stresses are below the ultimate strength of the metal

Diagram

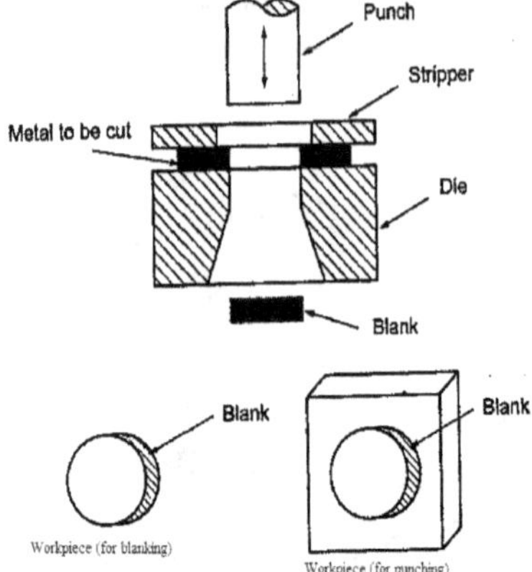

Figure 4.18

Blanking

- Cutting a flat shape from the metal
- The metal is punched out is called as blank

4.17. Punching

- Producing the hole on the work piece by a punch.
- The metal is removed is called scrap

Shearing

- Metal is cut along a single line

Diagram

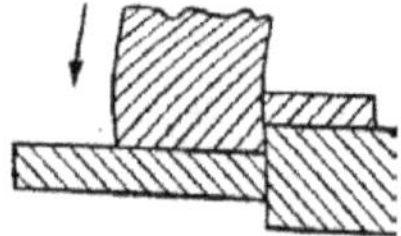

Figure 4.19

Parting

- Which the metal is cut simultaneously along two parallel lines

4.18. Notching

- Which metal pieces are cut from the edge of a sheet

Diagram

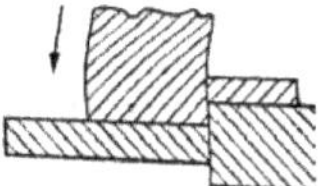

Figure 4.20

Trimming

- Used for removing excess metal

Shaving

- It is similar to trimming operation but here the amount of metal removal is usually about 10% thick of blank

Perforating

- Multiple holes which are very small

Sitting

- Unfinished cut through a limited length only

Diagram

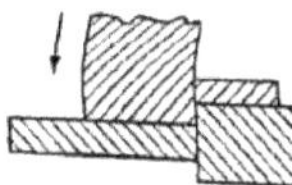

Figure 4.21

Launching

- Cutting the metal sheet through a small length

Diagram

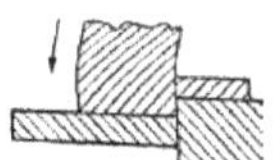

Figure 4.22

4.19. Forming Operation

Bending

- Forming the metal between a suitably shaped punch
- The included angle on the tool is usually smaller then be produced for allow spring back
- Spring back is a term which denotes the property of sheet metal to partially fall from its bending operation
- Punch forcing a sheet metal blank to flow availability between the punch

4.20. Squeezing

- The metal is caused to flow to all portions of a die capacity

4.21. Embossing

- Producing a required shapes on sheet metal

Diagram

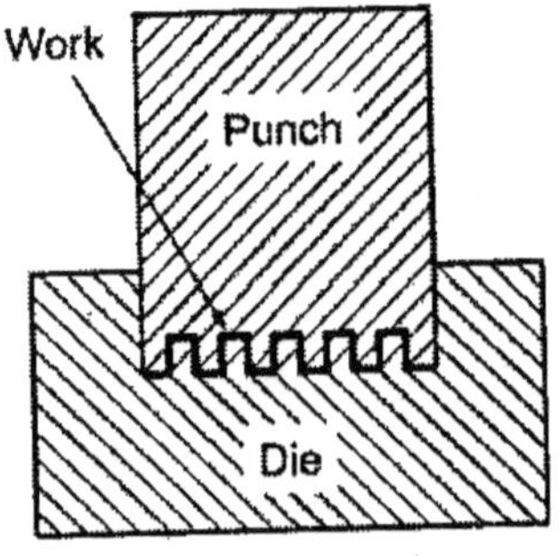

Figure 4.23

4.22. Nipping

- Cutting any shape from sheet metal without special tools
- The time taken to cut the required shape is less when compared to other cutting processes

Roll Bending

- The metal is bent at an angle to each edge

Diagram

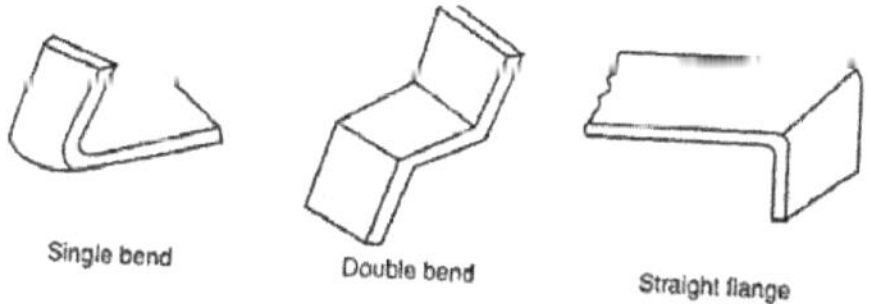

Figure 4.24

- If the metal is bent in the form of rolled edge at the edges of the work

Diagram

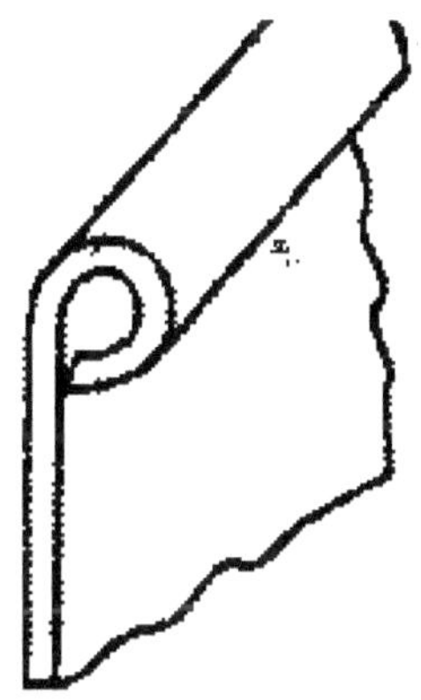

Figure 4.25

4.23. Roll Forming

- If the edges are formed to a desired shape

Diagram

Figure 4.26

4.24. Seaming

- The process of providing lock between the edges of the different work metal is called seaming

4.25. Stretch Forming Operation

Synopsis

- Introduction
- Diagram
- Working
- Merits
- De-merits
- Applications

Introduction

- Stretch forming is used for forming smoothly contoured
- The metal beyond the elastic limit to give the work piece a permanent set.

Diagram

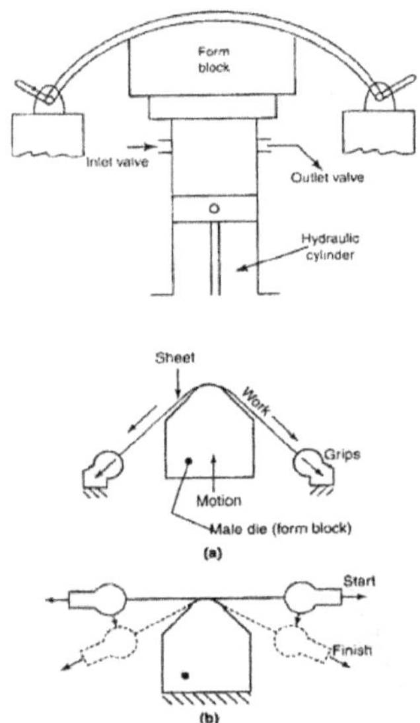

Figure 4.27: Mating Die Method

By Using Two Methods

- Form block method
- Mating die method

Form Blocks Method

- Two edges of the metals sheet are gripped are firmly
- The sheets is gripped on both sides by two jaws which are attached to the piston rod of hydraulic cylinders
- The piston and jaws can be pulled away and tensile forces are applied on sheet
- The cylinders can be moved in guide along a pre-determined path to stretch the sheet along the contour
- The jaws are mounted on two sides which are moved horizontally for providing tension to the sheet.
- Attached to the raw of hydraulic of press which moves upwards.

Diagram

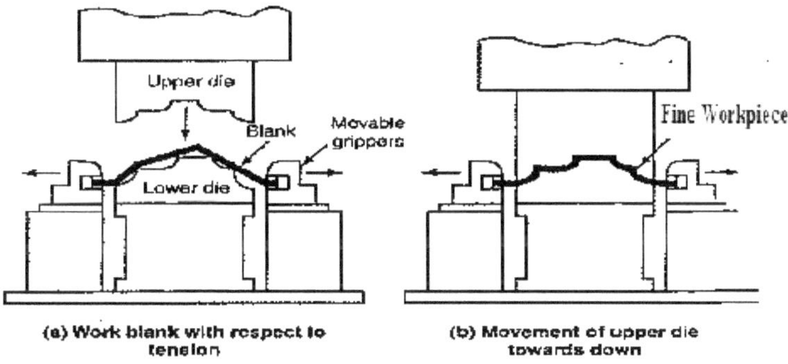

Figure 4.28

Merits

- No direct bending of the sheet
- Suitable for mass production
- Cost of tooling is low

De-merits

- It requires high quality from blocks
- Only simple and straight shapes can be stretched

- Applications
- Stretch forming is used in models of air craft, hood covers.

4.26. Formality of Sheet Metal

Synopsis

- Introduction
- Working
- Merits
- De-merits

Introduction

- It represents the response and responsibility of the material for forming process
- Material behave in diff manner at temperatures
- Some material may exhibit good formability at high temperature
- Some materials may readily flow when deformed at slow speed but sometimes they will break
- To find those limits which metal can be strained in any specific type of operation it has been developed.

Law 1: Process of Fracturing

- It states that, ductility of the metal is lower
- It refers to identical metal from which specimen diff section thickness have been machined and tested

Law 2: Geometrical Similitude

- Blanks which are geometrically similar each dimension width length of another blank can be fabricate
- Unit strains at corresponding location as geometrically
- Forces required to form geometrically similar blanks are directly proportional to square of thickness
- Find the temperature range for various specimen tests
- Test for bulk deformation
- Test for elastic plastic deformation
- Test for forming operations
- Scale forming test

4.27. Special Forming Process

Synopsis

- Introduction
- Diagram
- Working
- Merits
- De-merits
- Applications

Introduction

- Pressing the form over the sheet.
- The form tool is operated by hydraulic cylinder by using hydraulic fluid
- If the form tool is operated by any other method except hydraulic cylinder, in the forming process.
- Hydro forming rubber pad forming peen forming explosive forming magnetic pulse forming super plastic forming metal spinning

Hydro Forming Two Types

- Hydro mechanical forming
- Electro hydraulic forming

4.28. Hydro Mechanical Forming

- In this method the blank is placed on the draw ring then the dome is lowered. The punch is raised and pushed into the blank

Diagram

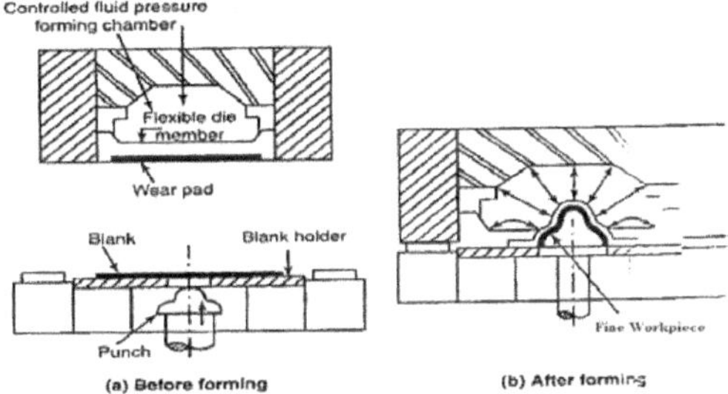

Figure 4.29: Hydro Mechanical Forming

Working

- The pressure in the pressure chamber contains increase at the end of cycle forming pressure is released forming chamber is raised.
- Then work piece is removed from punch

Merits

- Tooling cost is low
- Tolerance of 0.005mm is possible

4.29. Electro Hydraulic Forming

- Conversion of electrical energy into mechanical energy in liquid medium.

Diagram

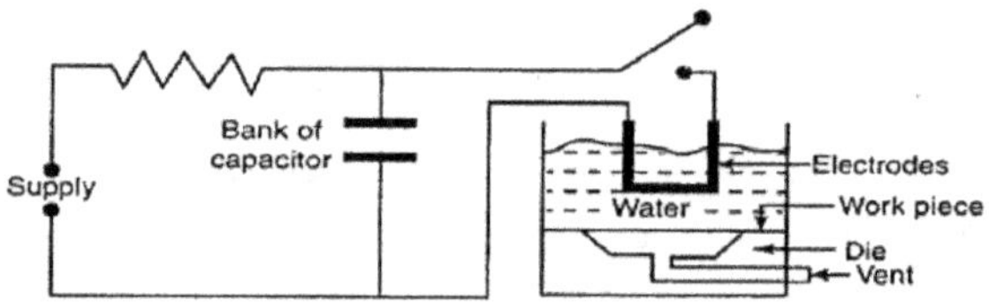

Figure 4.30: Electro Hydraulic Forming

Working

- Electric spark in a liquid produces shock waves and pressures can be used for metal forming
- During the process, high voltage electrical energy is discharged from a capacitor into a thin wire
- The whole unit is immersed into water vapour products start converting the electrical to hydraulic.
- The generated shock waves forces the metal

Merits

- Tooling cost is low
- High reproducibility

De merits

- The energy produced for forming is less
- This method is not suitable material having low ductility

Applications

- Aluminium bicycle frame
- Satellite antennas up to 6mm diameter

4.30. Explosive Forming

- It use of pressure wave generated an explosion in a fluid
- The explosive are used in form of sheet
  1. Standoff operation
  2. Contact operation

Standoff Operation

- The charge is located some distance away from the work piece and energy is transmitted through fluid
- Operating pressure for the work piece is between thousand to one lacks.

Contact Operation

- The charge is directly located over the blank
- This operation is mainly used for welding, Harding, compacting cutting process.

Diagram

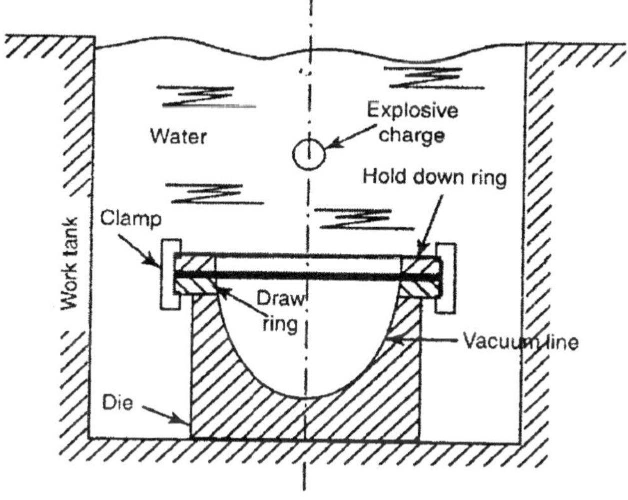

Figure 4.31

Working

- Charge is in direct contact with the work piece
- When explosive is detonated its energy is transmitted
- To avoid adiabatic compression .the space in the die behind the work piece generally evacuated
- If vacuum is not created between the work piece and die. Then an air cushion develop the metal is being forced into die
- When explosive detonated under water it produces shock waves.

Formulae

$P = Pme^{-t/0}$

Operations

- Blanking embossing coining drawing sizing

Merits

- Component is only formed in one cycle
- Large size component made easily
- Low capital investment required

De merits

- High skilled operators are required
- Suitable one for low quantity production

Application

- It's mainly used in aerospace industries

4.31. Rubber Pad Forming

Synopsis

- Introduction
- Diagram
- Working
- Merits
- De-merits
- Application

Introduction

- It also known as mar form process.
- It's a metal working process where sheet metal pressed between a die and rubber block
- Under pressure, the rubber and sheet metal are driven into the die.

Diagram

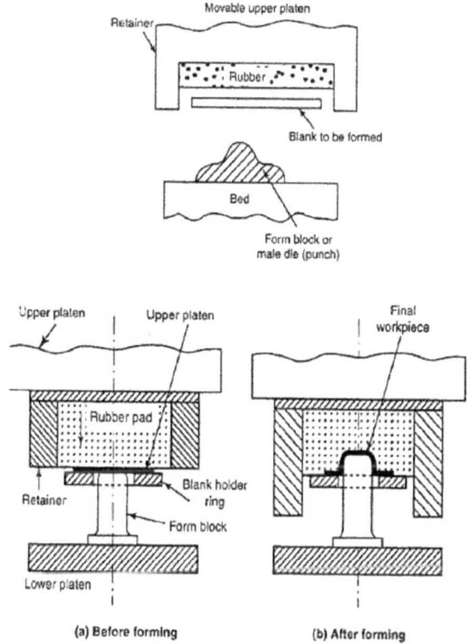

Figure 4.32: Rubber Pad Forming

Working

- It's used for bending and drawing operations
- In this method, punches is arranged at regular intervals along the pad
- Initially the blank is placed on the form block
- The force applied on the blank with the help of hydraulic cylinder through the ram
- The pad is generally made of rubber
- During forming the upper platen is moved to top surface of the blank. The required shape is obtained

* It is used to apply pressure on the blank.

Merits

* Cost of tooling is less
* Process is more flexible
* Lubricants are not required

De merits

* There is difficulty in the forming of shape
* It will wear out

Application

* It's used for unsymmetrical shape

4.32. Magnetic Pulse Forming

Synopsis

* Introduction
* Diagram
* Working
* Merits
* De-merits
* Applications

Introduction

* Apply a powerful magnetic pulse on metallic work piece

Diagram

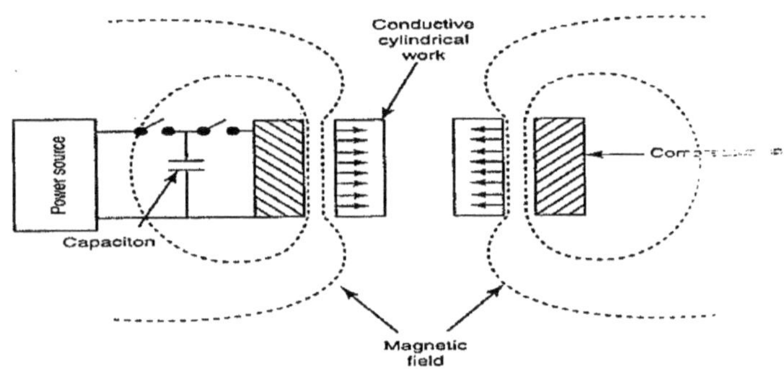

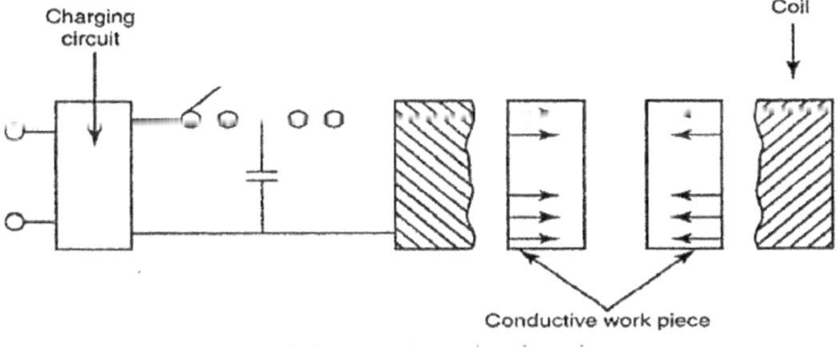

Figure 4.33: Magnetic Pulse Forming

Working

- Insulated induction is either wrapped
- The coil is shaped to produce the work piece and capacitor
- Capacitor can be charged slowly generally 3 to 6 sec
- When very high currents for a short time are passed through the coil, a magnetic field develops the work piece to collapse
- Energy storage ability of unit determines the size of work piece formed

Merits

- No friction, lubricants are not required
- No tool marks a work piece.

De merits

- Difficulty in forming of unsymmetrical shapes

Applications

- It's used attachment of rubber boots
- It's also used for shearing, sizing.

4.33. Peen Forming

Synopsis

- Introduction
- Diagram
- Working

Introduction

Throwing a blast of metal shots on to surface.

Diagram

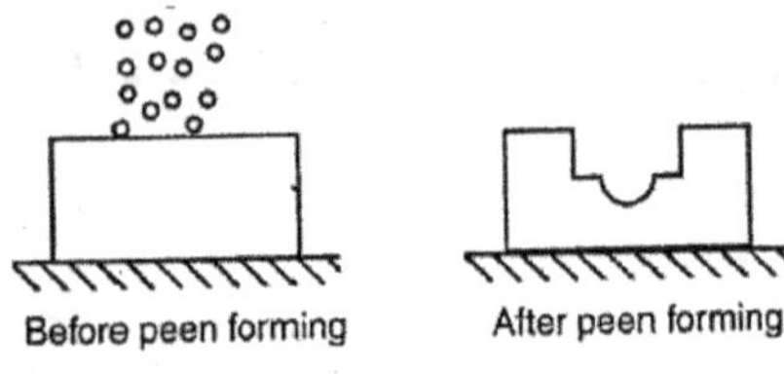

Figure 4.34: Peen Forming

Working

- The blast may be thrown either by using air pressure.
- The high velocity metal blast shot provides a sort of compression over the surface of component
- It also performed to prevent the cracking of work piece in corrosive media.
- The shots are made of cast iron.
- Efficiency of process mainly depends on the angle between the shots
- It generally used for coil springs, leaf springs.

4.34. Super Plastic Forming

Synopsis

- Introduction
- Diagram
- Merits
- De-merits
- Applications

Introduction

- Its metal working process used for forming sheet metal.
- In this metal defined by very high tensile ranging
- Components are formed by applying gas pressure.
- Process
- During this process the material is heated to the SPF temperature
- For titanium is around 900^0c
- Inert gas pressure is then applied.

- The flow stress of the material to take the shape of die pattern with increasing strain rate.
- Super plastic alloys can be stretched at high temperature by several times of initial length without breaking
- One of the material developed for super plastic
    - Zinc aluminium
    - Titanium
    - Aluminium

Diagram

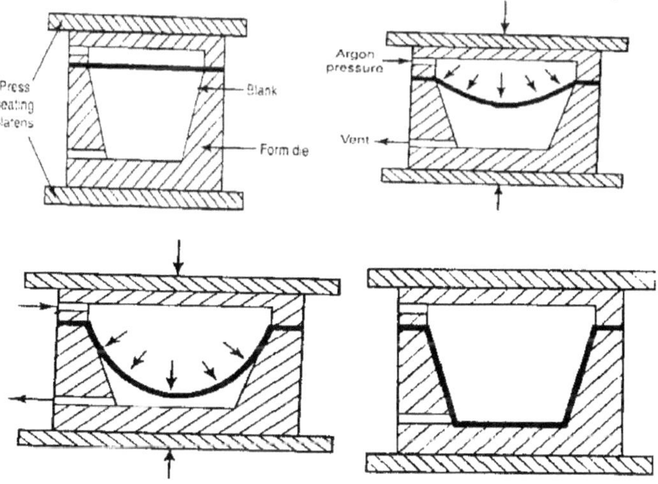

Figure 4.35: Super Plastic Forming

Merits

- Its eliminates unnecessary joints and rivets
- Less tooling cost

De merits

- Forming rate is low
- Cycle time varying from two minutes to two hours.

Applications

- Automotive body panels
- Aircraft frames

4.35. Metal Spinning

Synopsis

- Introduction
- Diagram
- Explanation
- Merits

Introduction

Pressure forming of metal on rotating chuck.

- It's used for making cup shapes.
- Spinning may be hot working.

Diagram

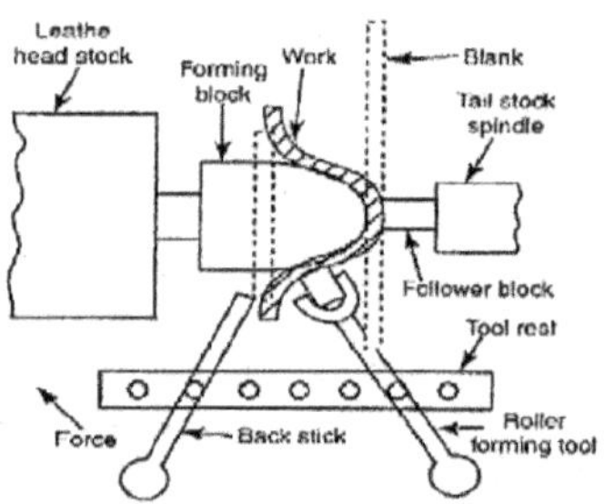

Figure 4.36: Metal Spinning

Explanation

- During the process a rotating disc of work sheet material is formed over a former
- It's applying pressure to outside of the disc by using forming tool.
- Generally applied in thin materials
- A spinning lathe is used
- The machine consists of bed, headstock, former
- The head stock of the machine, a hard metal form black shape of part is fixed
- After climbing the blank is rotated and metallic tool is moved back.

Merits

- Low equipment cost
- Low tooling cost.

UNIT 4

SHEET METAL PROCESSES

PART-A (2 MARKS)

1. What is blanking?
2. What is punching operation?
3. What are the different types of metals used in sheet metal work?
4. Mention any four products produced by spinning process?
5. In which member the clearance should be given for blanking and piercing?
6. What is the difference between stretch forming and bending?
7. List various operations generally performed in a sheet metal shop
8. Show the details of punching process with the help of a simple sketch
9. Give the difference between punching and blanking
10. List the various sheet metal that can be formed in press working
11. Define the term spring back
12. What are the advantages of stretch forming operation?
13. What are the types of special forming processes?
14. What are the advantages of hydro forming process?
15. State the limitations and applications of rubber pad forming process
16. What is metal spinning process?
17. State the advantages and applications of explosive forming process
18. What is peen forming process?
19. What are the advantages and disadvantages of peen forming process
20. What are the applications of super plastic forming process?

PART-B (16 MARKS)

1. i. Explain any one stretch forming operation

   ii. Define formability and how it is tested?

   iii. What is drawing operation?

2. i. Explain the metal spinning operation

   ii. Describe the magnetic pulse forming process

3. What is deep drawing operation? Explain with a neat sketch.

4. i. Explain rubber pad forming process

   ii. Describe the electro hydraulic forming process

5. i. Describe the explosive forming process

ii. How are aluminium kitchen utensils produced?

6. i. Describe the process of hydro forming

ii. Describe the various methods of rubber forming. Where are these processes used?

7. i. What is super plastic forming?

ii. Describe the hydro forming process with the help of neat diagram

8. i. Explain the characteristic features of sheet metal used in forming process

ii. Explain peen forming process

9. i. Find the total pressure, dimensions of tools to produce a washer 5cm outside dia with a 2.4 cm diameter hole, from a material 4 mm thick, having a shear strength of 360 N/mm^2

ii. Determine a) blank diameter b) Least no. drawing operations c) force and energy for the first draw with 40% reduction to produce a cup of 5 cm in diameter and 7.5 cm deep to be drawn from 1.5 mm thick drawing steel with a tensile strength of 315 N/mm^2

PLASTICS & POLYMERIZATION

5.1. Introduction

- Many organic materials used in engineering industries Organic materials are those materials obtained from carbon, and chemically combined with other non-metallic compounds

Two Type

- Nature organic materials
- Synthetic organic materials

Nature Organic Materials

- The wood, coal petroleum

Synthetic Materials

- The plastic synthetic rubbers ceramic glasses

Polymers

- It's a Greek technology
- Polymers means many means part
- The single unit is called as monomer
- It's a small molecules thousand monomer joined together to form a large molecule

Polymerisation Process

- Addition polymerisation
- Condensation polymerisation

Addition Polymerisation

- Large numbers are added chemically one by one
- The monomers form a long chain molecule
- The bonding is Vander wall force
- In this process no catalyst is added
- The addition two monomers is called co polymerisation

Condensation Polymerisation

- Two or more unlike monomers are linked and smaller molecules from by a product
- During this process by product such as water is formed

- Materials used for processing of plastics

Additives

- To improve the plastic behaviour
- It acts as an internal lubricant for increasing the toughness and flexibility

Example

- water organics

Catalyst

- Added to the promote faster

Dyes and Pigments

- Added to import a desired colour

Imitators

- Initiate the reaction
- Allows to begin polymerisation

Modifiers

- Used to improve the mechanical properties

Lubricants

- It's used to reduction friction during processing to prevent parts from sticking to mould walls

Flame Retardants

- Added to the plastic to enhance the non in flammability of plastics

5.2.   Properties of Plastics

- Elongation
- Heat resistance
- High rigidity
- High viscosity
- Surface hardness
- Density
- Ignition temperature
- General chemical resistance

5.3. Types of Plastics

- Thermosetting plastics
- Thermo plastics

Thermosetting Plastics

- Which are hardened by heat effecting a non-reversible chemical change are called thermosetting
- Do not soften on re heating
- It's are formed by condensation polymerisation
- Very strong binding force between molecules

Phenol Formal Eyed

- It's also named as Bakelite
- It made by the reaction of phenol with formaldehyde
- Generally produced in dark colour
- It has high strength, stability

Applications

- Pulleys, bottle caps ,tooling

Polyester Resin

- It has low moisture good electrical resistance
- It's used in paper mat, TV parts and car bodies
- The main drawback of the polyester is high cost

Melamine

- It has excellent electrical and heat resistance
- The melamine's are available Catlin Plasmon
- Its widely used for moulded parts

Phenol Furfural

- It has good flow at low moulding temperatures

Example

- Brake linings, electrical parts

Uses

- It's used as a binder in retinoid

Epoxy Resins

- The most popular variety of epoxy resins is araldite
- It has good electrical and chemical resistances

Uses

- Tools and dies jigs and fixtures

Silicones

- It have high resistance to high temperature up to 260^0C and possess excellent dielectric strength at temperatures

Uses

- It's used for coatings ,laminates, foam products

Urea Formaldehyde

- It is obtained by the condensation of urea and aqueous formaldehyde

Uses

- It's used in toilet seats, table ware, buttons

Alkyds

- It also known as oil modified polyesters. Alkyds are used in synthetic enamels and lacquers. It is used in solid form where high electrical and heat resistance are required.

Example

- Automobile ignition parts.

Polyurethanes

- It is mainly used for cushions in transportation seats for insulation and electronic equipment as a packing material.

Thermoplastics

- The thermoplastics have separate long and large size molecules arranged side by side.
- Some of the   thermo plastic structure is amorphous in nature other than that all the crystalline structure in nature.
- When thermo plastics are heated, it becomes very soft and re-hardens on cooling.
- It is easily remoulded or extruded to any shape.
- These plastics do not have a definite temperature.
- The various thermo plastics are discussed below.

Cellulose Derivatives

Synthetic Resins

- Cellulose derivatives

Cellulose Nitrate

- It is obtained by treating the cellulose with a mixture of nitric and sulphuric acid.
- It has high toughness.

Used

- Spectacle frames, toilet articles

Cellulose Acetate

- Its obtained by treating the cellulose with acetic acid
- Its lighter then cellulose

Uses

- Photographic films
- Buttons radio panels

Ethyl Cellulose

- Lightest of all cellulose derivatives
- It has good electrical properties

Uses

- Jigs and fixtures
- Hose nozzles

Cellulose Acetate –Butyrate

- Its obtained by treating cellulose with acetic acid and butyric acid
- It has good stability

Cellophane

- Its available in extruded form
- It has good resistance

Uses

- Curtains drapers wrapping

Cellulose Propionate

- It has low tendency for moisture absorption
- It can be easily moulded

Uses

- Fountain pens telephones

Polyethylene

- It has very high resistance to acids solvents can be flexible
- Tough and good insulators
- It has low water absorption

Uses

- Fabrics trays pipes and tubing chemical containers

Polystyrenes

- It has dimensional stabilities and strain resistance
- It easily mouldable

Uses

- Gaskets
- Greaseless bearing

Polyamide

- Its popularly known by its trade by name nylon
- It has high strength and elasticity
- It can be moulded into rods

Uses

- Yarn for cloth
- Bearing and couplings

Methyl Methacrylate

- It trade name is Lucite and Plexiglas
- It can be formed easily at temperatures around 120^0c
- Its marked by its colour
- High transmission capability

Tabulation 5.1: Difference between Thermo Plastics and Thermo Setting Plastics

S.No	Thermo plastics	Thermosetting plastics
1	Its softened by heating	Cannot be softened by this process
2	Structure is made of linear chain molecules	Structure is made of linked molecules
3	Produced by addition polymerization process	Produced by condensation polymerization
4	It can be re produced by heating and cooling	It cannot be re produced
5	The temperature increases with increase in plasticity	Plasticity is stable at high temperatures
6	It can be remoulded to any shape	It cannot be remoulded
7	Its softer and less strong	It's harder and strong
8	Scrub can be used	Scrub cannot be reused

Tabulation 5.2: Moulding

S.No	Processes	Characteristics
1	Injection moulding	high production rates Good dimensional accuracy
2	Blow moulding	Hollow thin walled parts Low cost for making container
3.	Rotational Moulding	Low tooling costs Low production rates
4.	Extrusion	Wide tolerance High production rates Low tooling costs
5.	Thermo forming	Medium production rates Shallow deep cavities
6.	Compression moulding	Relatively inexpensive tooling Medium production rates
7.	Transfer moulding	Some scrap loss Medium tooling cost High production rates

5.4.   Forming and Shaping of Plastic

Synopsis

- Introduction
- Diagram
- Working

Introduction

- Injection moulding is used achieve high speed moulding of thermo plastics.
- Molten thermo plastic is injected into a mould under high pressure.

Two Types

- Ram type
- Screw type

Ram Type Injection Moulding

- Injection unit
- Climbing unit

Diagram

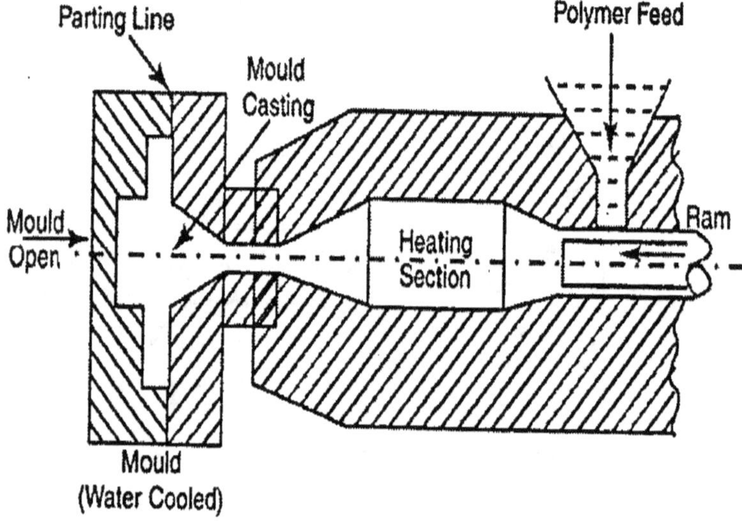

Figure 5.1: Ram Type Injection Moulding

Working

- The polymer is filled in hopper.
- Heating section the polymer is melted and the pressure is increased.
- The heated material is injected by ram under pressure
- Heated material is forced to fill in the mould cavity through the nozzle to get required shape.

5.5.   Screw Type Injection Moulding

Two Units

- Injection unit
- Clamping unit

Diagram

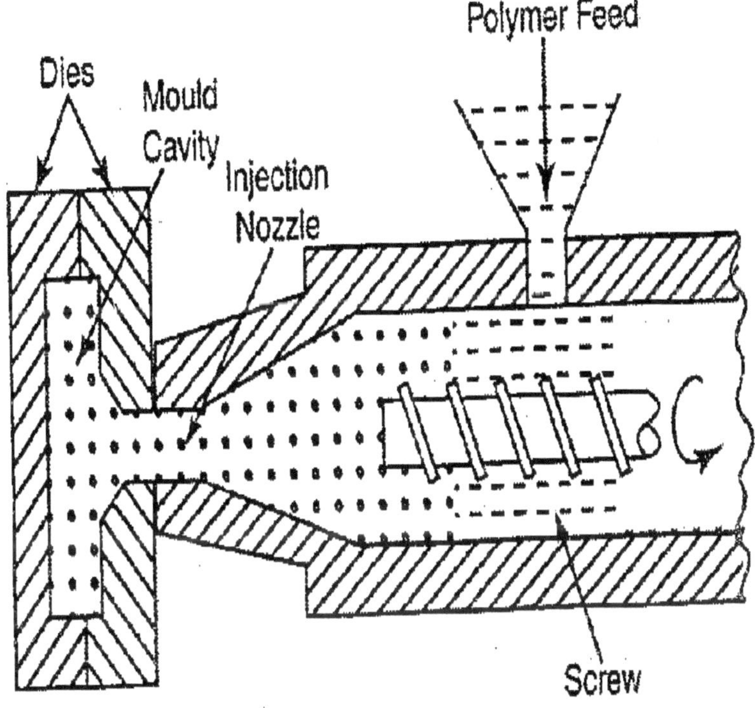

Figure 5.2

Working

- The injection unit has hopper.
- In this machine the pellets are initially fed into the hopper.
- The resins are pushed along with the heated screw.
- The screw itself moving down wards accumulation allocation of enough material to fill the mould.
- The rotation of the screw provides the plastering action.
- Low viscosity of monomers is used.
- A chemical reaction takes place between at resins at low temperature.
- The plastic is pre heated in 93^0c
- The injection capacity of range 12000 mm

Merits

- The cost is low
- Wide range of shapes can be moulded.

De merits

- It's used making parts of complex threads
- Indicate shapes such as thin wall parts can be produced.

Applications

- The temperature controls are essential.

5.6.   Blow Moulding

Synopsis

- Introduction
- Diagram
- Working
- Merits
- De-merits
- Applications

Introduction

- A hot extruded tube of plastic called parison is placed between two parts of open moulds.

Diagram

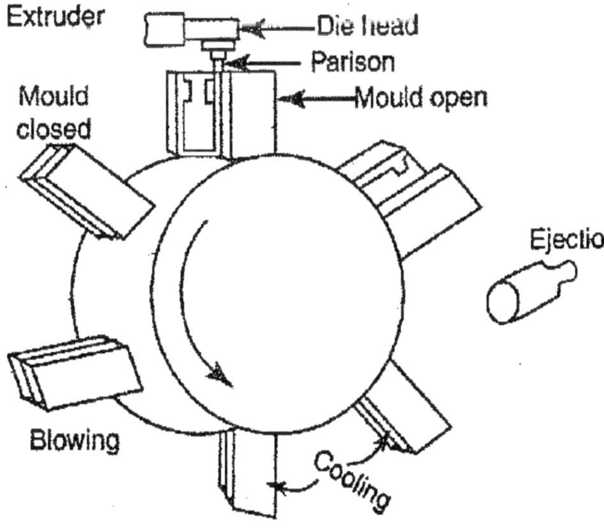

Figure 5.3: Blow Moulding

Working

- The bottom end of parison is scaled.
- The compressor air is used to blow the molten plastic into the mould.
- The air pressure is about 0.7 to 10 kg/cm².
- The component is cooled and the mould opens to release the components.

Merits

- The cost is low
- Extensive of profiles can be moulded.

De merits

- It's used making parts of complex gears
- Indicate shapes such as thin wall parts can be produced

Applications

- It is used in making plastic bottles & toys.
- The hollow containers are produced by this process.

5.7.   Rotational Moulding

Synopsis

- Introduction
- Working
- Uses

Introduction

- The rotational moulding process is used make thin walled hollow parts.
- In this method a measured quantity of polymer powder is placed in thin walled metal mould.

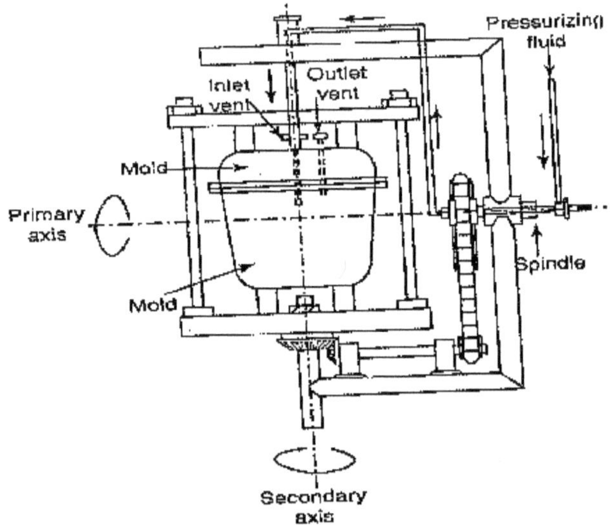

Figure 5.4: Rotational Moulding

Working

- In this rotational moulding, thin walled metal mould is made of two pieces is rotated in perpendicular axis.
- Plastic material is placed inside the mould.
- The mould is heated and rotated.
- Most thermoplastics and some thermo sets can be formed into large hollow parts by rotational moulding.

Uses

- It is used to produce toys in P.V.C.
- It is used to make large containers of polyethylene.

5.8. Film Blowing

Synopsis

- Introduction
- Diagram
- Working

Introduction

- Crystalline sharp melting polymers such as nylon or PET are very much suited for the film productions by melt casting and techniques.

Diagram

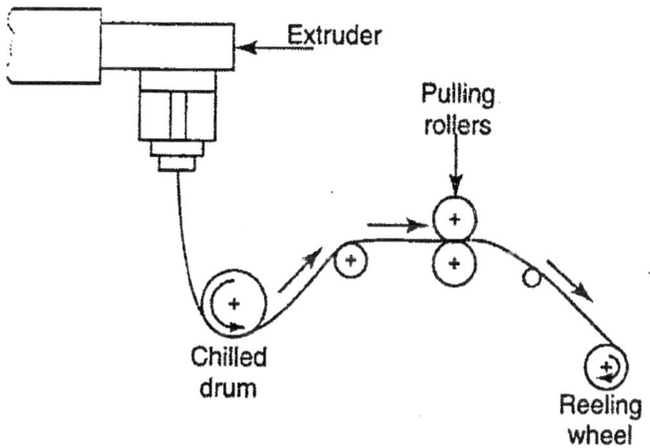

Figure 5.5: Flim Blowing

Working

- The film is produced.
- Thin film it is stretched by pulling rollers through the chilled drum in the reeling wheel.
- The reeling wheel is used to make the film roll.

5.9. Sheet Making

Synopsis

- Introduction
- Diagram
- Working
- Applications

Introduction

- Calendaring process is used for sheet making
- The plastic compound at resin filter and other additives heated for some times and passed through the rollers.
- It's similar to rolling process.

Diagram

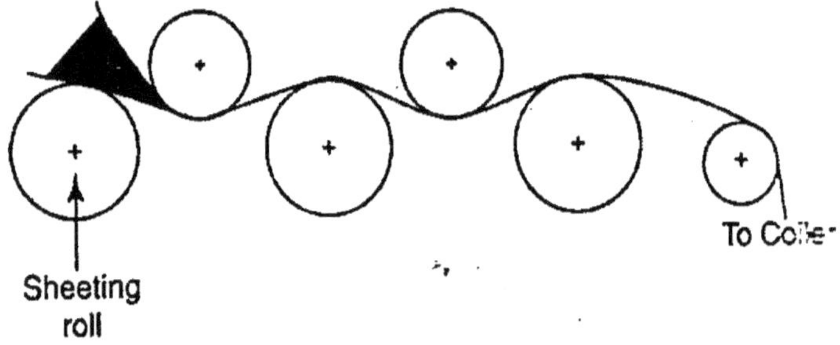

Figure 5.6: Sheet Making

Working

- The thickness of the sheet is controlled by combination altering speed of rolls. Thickness determined gab between rolls at final stage.
- In these process 1st role-feeder 2nd-metering device 3rd-gaps sets the gauge Wound in a coiler

Applications

- It's used for making PVC tubes.
- It's also used for making floor tiles.

5.10. Vacuum Forming Process

Synopsis

- Introduction
- Diagram
- Working

Introduction

- It process heated plastic sheet changed to desire shape.

Diagram

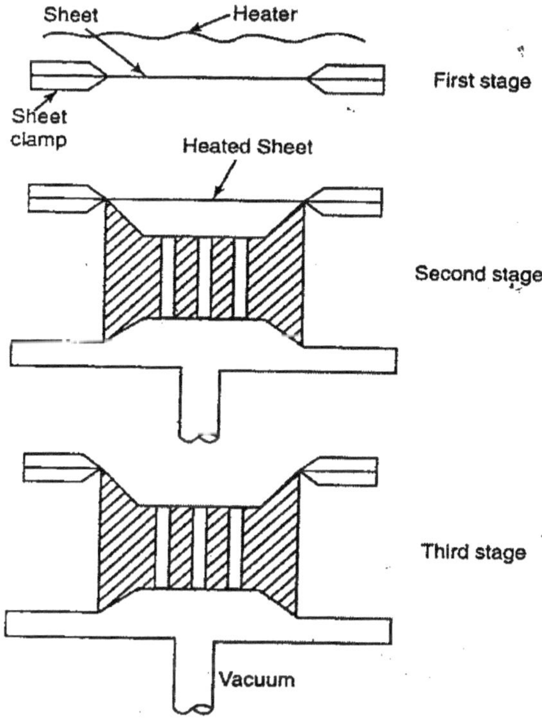

Figure 5.7: Vaccum Forming Process

Working

- The mould surface by reducing the air pressure between one side of sheet and mould surface.

1st Stage

- The plastic sheet is heated in a heater and the sheet is fixed in clamp

2nd Stage

- The heated sheet is placed on the die where air between the sheet and mould is removed.

3rd Stage

- Increasing intensity draws the sheet against the surface of mould, where it cools
- The vacuum forming process are also called thermo forming
- The mechanical assist is given to stretch plastic into the mould

5.11. Ultrasonic Welding

Synopsis

- Introduction
- Diagram
- Working
- Uses

Introduction

- Its conversion of high frequency electrical energy into high frequency electrical energy.
- The mechanical energy vertical motion excess of 15000 cycles/sec

Diagram

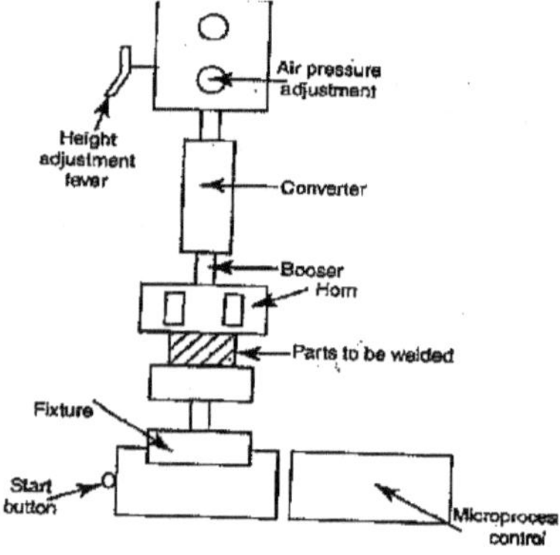

Figure 5.8: Ultrasonic Welding

Working

- The vibration motion transferred to thermo plastic pressure.
- Frictional heat is generated at the inter face. It's done through an ultrasonic welder.
- The welding power supply converts standard 60 Hz alternating to 20000 Hz
- The alternating current enters the converter where it's converted to vertical mechanical motion equal to 20000 Hz
- Then it's passed through a booster which can increase the amplitude of vibrating motion.

Uses

- It's used for assembling thermo plastic material
- It's used in automotive, toy production.

5.12. Induction Welding

Synopsis

- Introduction
- Diagram
- Explanation
- Uses

Introduction

- In this process induction welding is used

Diagram

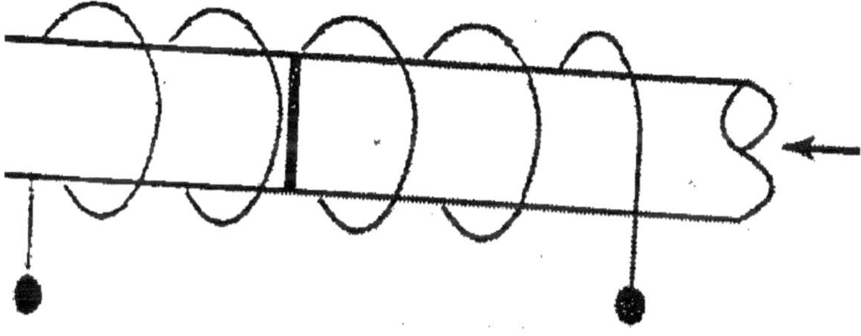

Figure 5.9: Induction Welding

Explanation

- In this process a high frequency generator produces an oscillating magnetic field with in a coil that close with the work piece
- The weld material is affected by this magnetic field and it generates heat itself
- Cavity conducting heat to the host material.
- So the molten motel forming a high integrity joins when fused together
- The weld material is directly with in the joint.

Uses

- Induction process is fast.
- It does not produce an external weld.

The Heading Tool Comes Back and the Movable Split Die Releases the Stock

Upsetting machines, called upsetters, are generally horizontal acting. When designing parts for upset – forging, the following three rules must be followed.

1. The length of unsupported bar that can be upset in one blow of heading tool should not exceed 3 times the diameter of bar. Otherwise bucking will occur.
2. For upsetting length of stock greater than 3 times the diameter the cavity diameter must not exceed 1.5 times the dia of bar.
3. For upsetting length of stock greater than 3 times the diameter and when the diameter of the upset is less than 1.5 times the diameter of the bar, the length of un – supported stock beyond the face of die must not exceed diameter of the stock.

The Heading Tool Comes Back and the Movable Split Die Releases the Stock

Upsetting machines, called upsetters, are generally horizontal acting. When designing parts for upset–forging.

The following three rules must be followed.

1. The length of unsupported bar that can be upset in one blow of heading tool should not exceed 3 times the diameter of bar. Otherwise bucking will occur.
2. For upsetting length of stock greater than 3 times the diameter the cavity diameter must not exceed 1.5 times the dia of bar.
3. For upsetting length of stock greater than 3 times the diameter and when the diameter of the upset is less than 1.5 times the diameter of the bar, the length of un–supported stock beyond the face of die must not exceed diameter of the stock.

The Heading Tool Comes Back and the Movable Split Die Releases the Stock

Upsetting machines, called upsetters, are generally horizontal acting. When designing parts for upset-forging, the following three rules must be followed.

1. The length of unsupported bar that can be upset in one blow of heading tool should not exceed 3 times the diameter of bar. Otherwise bucking will occur.
2. For upsetting length of stock greater than 3 times the diameter the cavity diameter must not exceed 1.5 times the dia of bar.
3. For upsetting length of stock greater than 3 times the diameter and when the diameter of the upset is less than 1.5 times the diameter of the bar, the length of un – supported stock beyond the face of die must not exceed diameter of the stock.

The Heading Tool Comes Back and the Movable Split Die Releases the Stock

Upsetting machines, called upsetters, are generally horizontal acting. When designing parts for upset – forging, the following three rules must be followed.

1. The length of unsupported bar that can be upset in one blow of heading tool should not exceed 3 times the diameter of bar. Otherwise bucking will occur.
2. For upsetting length of stock greater than 3 times the diameter the cavity diameter must not exceed 1.5 times the dia of bar.
3. For upsetting length of stock greater than 3 times the diameter and when the diameter of the upset is less than 1.5 times the diameter of the bar, the length of un–supported stock beyond the face of die must not exceed diameter of the stock.

UNIT 5

MANUFACTURING OF PLASTIC COMPONENTS

PART-A (2 MARKS)

1. What are the applications of laminated plastics?
2. Enumerate the advantages of Electro-hydraulic forming process
3. Write any two limitations of Electro-magnetic forming process
4. What are the characteristics of thermosetting plastics?
5. Classify thermoplastics
6. What is meant by high energy rate forming?
7. Name any four thermosetting plastics, used in industries.
8. Give four examples of thermo plastics
9. Give two examples of thermoplastics and thermosetting plastics
10. What is blow moulding process?
11. Which type of plastic is used for manufacturing electrical switches? Why?
12. What is meant by rotational moulding?
13. List the advantages and disadvantages of rotational moulding
14. Define thermoforming
15. What is meant by transfer moulding?
16. Which type of moulding is used for making bottles?
17. What are the types of compression moulding?
18. What is bonding of thermoplastics?
19. List the advantages and disadvantages of transfer moulding
20. What is meant by film blowing?

PART – B (16 MARKS)

1. Explain the principle of injection moulding process
2. i. Describe any method of bonding thermoplastics

   ii. What is laminating? Explain the low pressure method of laminating
3. i. Explain the transfer moulding process

   ii. Why screw injection moulding machine is better than a ram type injection Moulding machine?
4. i. Describe the compression moulding process

   ii. Describe briefly any two thermoplastics and thermosetting plastics

5. What are the processes used for processing of thermoplastic? Explain any one process with suitable sketches

6. What is thermoforming process? Explain with a neat sketch

7. Describe film blowing operation

8. Explain Rotational moulding

9. i. Explain blow moulding process with its salient features

    ii. What are the additives to be mixed in processing plastics and explain the purpose of each?

10. i. Describe different types of plastics with applications of each type

    ii How do thermoplastics differ from thermosetting plastics?

QUESTION PAPER CODE : 80654

B.E./B.Tech. Degree Examination, November/December 2016.

Fifth Semester

Mechanical Engineering (Sandwich)

ME6302 – Manufacturing Technology – I

(Common to Third Semester Industrial Engineering, Industrial Engineering and Management, Mechanical and Automation Engineering and Mechanical Engineering)

(Regulations 2013)

Time: Three hours   Maximum: 100 marks

Answer All Questions

Part – (10 x 2 = 20 marks)

1. What are the characteristics of a core?

2. Name the alloys which are generally die cast. Why is aluminium alloys preferably cast in cold chamber die casting machines?

3. Why do residual stresses get developed in weldments?

4. Why the temperature in plasma arc welding is much higher than in other arc welding processes?

5. Why is it necessary to condition the metal before hot rolling?

6. Give a few examples of hot forged products.

7. What are the desirable qualities in metal for maximum stretchability?

8. What are the applications of rubber pad forming process?

9. Name the various methods of processing thermoplastics

10. Define film blowing.

Part B–(5 x 13 = 65 Marks)

11. (a)(i) How are patterns classified? Describe any two types with sketches and State the uses of each of them.   (7)

    (ii) Enumerate the casting defects and suggest suitable remedies   (6)

OR

(b) (i) Explain the process of centrifugal casting with suitable sketch and state its specific applications   (8)

(ii) What are the main characteristics of mould sand?   (5)

12. (a) (i) Compare MIG and TIG welding in respect of their principle of working and field of application   (8)

(ii) What is a soldering flux? What different types of soldering fluxes are used? (5)

OR

(b) Write short notes on (i) Electron beam welding (ii) Friction stir welding   (6+7)

13. (a) (i) Briefly explain the various operations performed in forging process   (7)

(ii) With suitable sketches, explain the stages involved in Shape rolling of structural sections.   (6)

OR

(b) (i) Explain the working of Mannesmann process with neat sketch   (7)

(ii) How is tube drawing carried out? Explain with suitable sketch   (6)

14. (a) (i) Explain the various sheet metal forming operations with neat sketches (8)

(ii) Discuss with neat sketch, the working of metal spinning process   (5)

OR

15. (a) Describe the following plastic processing methods with neat Sketches

(i) Compression moulding (ii) Blow moulding   (7+6)

OR

(b) (i) Why is the thermoforming a valuable method for the plastic Manufacturer? Explain the process with neat sketch   (7)

(ii) State the purpose of the following in plastics (1) Plasticizers; (2) Fillers and (3) Stabilizer.   (6)

PART C – (1 X 15 = 15)

16. (a) Derive the mathematical expression for the Flat strip metal rolling process to calculate the rolling load.   (15)

OR

(b) A casting is required to have the following composition: C-3.25%, Si-1.8%, Mn-0.6%, P-0.5% and S-0.1%. Determine the weight of pig iron from pile A and Pile B to be picked up in each metal charge if the charge (200 kg) is to contain pig iron–50%, foundry return–40% and purchased scrap–10%. Analysis of these metals is as follows:

Metal	Si%	Mn%	S%	P%
Pig iron (pile A)	2.4	0.9	0.05	0.4
Pig iron (pile B)	1.4	0.55	0.05	0.33
Foundry returns	1.7	0.6	0.06	0.3
Purchased scrap	2.2	0.7	0.07	0.25

www.ingramcontent.com/pod-product-compliance
Lightning Source LLC
LaVergne TN
LVHW022051190726
843495LV00014B/1741